Student Guide and Workbook

To Accompany

Biochemistry and Molecular Biology

Student Guide and Workbook

To Accompany
Biochemistry and Molecular Biology
by William H. Elliott and Daphne C. Elliott

John R. Jefferson
Associate Professor of Chemistry, Luther College, Decorah, Iowa, USA

Oxford • New York • Melbourne
OXFORD UNIVERSITY PRESS
1998

Oxford University Press, Great Clarendon Street, Oxford OX2 6DP

Oxford New York

Athens Auckland Bangkok Bogota Bombay
Buenos Aires Calcutta Cape Town Dar es Salaam
Delhi Florence Hong Kong Istanbul Karachi
Kuala Lumpur Madras Madrid Melbourne
Mexico City Nairobi Paris Singapore
Taipei Tokyo Toronto Warsaw

and associated companies in
Berlin Ibadan

Oxford is a trade mark of Oxford University Press

Published in the United States by
Oxford University Press Inc., New York

A catalogue record for this book is available from the British Library

Library of Congress Cataloging in Publication Data
(Data available)
ISBN 0 19 857812 1

Typeset by the author
Printed in Great Britain by
The Bath Press

Die Arbeit macht das Leben süss.

(Work makes life sweet)

old German saying

Few things in life are as satisfying as a good idea.

JRJ

Preface

....

This companion text has been written to complement the main text, *Biochemistry and molecular biology* by William H. and Daphne C. Elliott. The intent is to be as 'student-friendly' as possible and this guide and workbook is meant to help in the learning process. Use your lecture notes to define which parts of the main text are to be emphasized since the individual course instructor is the ultimate authority with respect to what must be known for the exam.

One of the main strengths of the text by Elliott and Elliott is that the chapter topics are arranged according to a progression of biological function. This makes it possible to view the biological context of the chemical systems in combination, which is the essence of biochemistry and molecular biology. Each of the chapters in the *Student guide and workbook* is designed to complement the corresponding chapter in the main text.

The following is a list of the different teaching and learning tools found in this companion text.

- There is a brief **Chapter summary**, which gives an overview of the highlighted topics.
- The checklist of **Learning objectives** can be used by the student to keep track of their progress.
- The **Walk through the chapter** follows the topical sequence of the material in the text and gives the student an opportunity to practise using appropriate terminology by filling in the blanks provided. The answers for a given section are found directly below it so that the student can find them quickly without having to flip to the back of the book.
- **Figure** reproductions from the main text have been incorporated in this workbook; key terms in each figure have been replaced by blank boxes to be filled in by the student. The correct answers will be found in the original figures of the main text.
- There is a **Review of problems from the end of the chapter** of the main text which are intended to give a more complete answer along with necessary background information.
- Finally there are **Additional questions** for each chapter with answers found in a section at the end of the book.

In addition to these teaching approaches included for each of the chapters of the main text, the *Student guide and workbook* also includes sections entitled 'Useful study tips before you begin' and 'Review of some principles of organic chemistry'.

Acknowledgements

For their patience, I would like to thank the JRJs—Jennifer, Joseph, Julie, Jason, James, and Jillian–and Carol because she is special. For their expert typing skills I would like to thank Lynn Williams and Janine Brandt. And finally, I would like to thank the staff at Oxford University Press for all of their indispensable skills.

Decorah, Iowa   J.R.J.

Contents

··

Useful study tips before you begin

One great key to success is to arrange your work load and prepare a method of working. The list below is meant to help you master some of the basics of good study techniques, problem-solving approaches, and examination-taking.

1. Get to know your textbook.

    - Write your name on the inside of the front cover.
    - Use the table of contents as a checklist for the things you have covered.
    - Look at the list of abbreviations periodically and keep track of the number of the chapter in which each was introduced.
    - Locate the answers to the problems at the end of the book.
    - The main text has an index of key words with page references.

2. Study with more than one book open.

    - Textbook.
    - Companion.
    - Lecture notes.
    - Dictionary.
    - Something very pleasurable and easy to read to be used to reward yourself after finishing a chapter.

3. Classify the types of knowledge you will be gaining and the best way to organize these.

    - Definitions.
    - Explanation and relationships.
    - Chemical equations, structures, and mechanisms.
    - Mathematical relationships.

4. Consider the following approach to each chapter.

    - At the beginning of each chapter, read through the brief summary paragraph in this guide.
    - As you read the main text, fill in the blanks and answer the questions in the 'A walk through the chapter' sections of the guide.
    - Try to do the exercises at the end of each chapter in the main text.

- Answer the additional questions in the guide.
- Use the 'Learning objectives' list to check off (√) the important concepts as you master them.

5. Keep track of which concepts in the chapter are emphasized by your instructor by marking the appropriate sections of the 'Learning objectives' to indicate their importance.

6. Review often. Set aside a specific time each day and keep to the schedule.

For more study tips see *Becoming a master student* by Dave Ellis, Houghton Mifflin Company (1997).

Review of some principles of organic chemistry

···

Names of classes of organic molecules based on functionality

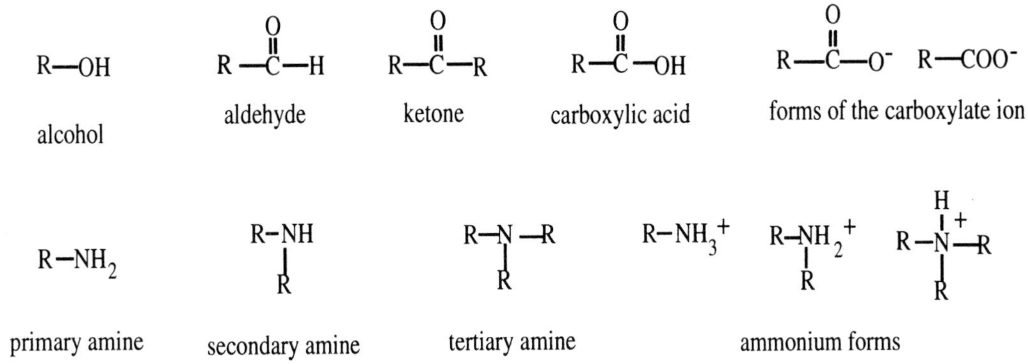

···

$$R-OH$$

alcohol

$$R-\overset{\displaystyle O}{\overset{\|}{C}}-H$$

aldehyde

$$R-\overset{\displaystyle O}{\overset{\|}{C}}-R$$

ketone

$$R-\overset{\displaystyle O}{\overset{\|}{C}}-OH$$

carboxylic acid

$$R-\overset{\displaystyle O}{\overset{\|}{C}}-O^- \quad R-COO^-$$

forms of the carboxylate ion

$$R-NH_2$$

primary amine

$$R-\underset{R}{\overset{\displaystyle}{N}H}$$

secondary amine

$$R-\underset{R}{\overset{\displaystyle}{N}}-R$$

tertiary amine

$$R-NH_3^+ \quad R-\underset{R}{\overset{\displaystyle}{N}H_2^+} \quad R-\underset{R}{\overset{\displaystyle H}{\overset{|}{N}}}-R$$

ammonium forms

···

Terms used to name functional groupings as attachments

Note: These terms are often used as suffixes and incorporated into the complete name.

....

—OH   —SH $\overset{\displaystyle O}{\overset{\|}{—C}}$—R $\overset{\displaystyle O}{\overset{\|}{—C}}$—CH$_3$ $\overset{\displaystyle O}{\overset{\|}{—C}}$—

hydroxyl   sulfhydryl   acyl   acetyl   carbonyl

—NH$_2$ —NH$_3{}^+$ $\overset{\displaystyle O}{—O—\overset{\|}{\underset{|}{P}}—O^-}$ $\overset{\displaystyle O}{—\overset{\|}{\underset{|}{P}}—O^-}$

amino   ammonium   O-   O-

                               phosphate   phosphoryl

....

There is some flexibility as to which portion of the molecule is considered the attached functional group and which is considered the parent, for example, phosphoryltyrosine is the same as tyrosinylphosphate.

....

....

Many bond types are made by condensing two functionalities

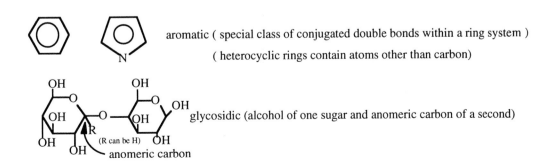

ester (from a carboxylic acid and an alcohol)

thio esters (from a carboxylic acid and a thiol)

amide (from a carboxylic acid and an amine)

phosphate ester (from an alcohol and a phosphate)

mixed phosphate anhydride (from a carboxylic acid and a phosphate)

diphosphate anhydride (from two phosphates)

disulfide (two thiols with an S-S bond)

trans double bond   cis double bond

aromatic (special class of conjugated double bonds within a ring system)
(heterocyclic rings contain atoms other than carbon)

glycosidic (alcohol of one sugar and anomeric carbon of a second)

(R can be H)

anomeric carbon

Some types of chemical reactions

$$R—OH \quad + \quad R'—\overset{\overset{\displaystyle O}{\|}}{C}—OH \quad \underset{\text{hydrolysis}}{\overset{\text{condensation}}{\rightleftharpoons}} \quad R'—\overset{\overset{\displaystyle O}{\|}}{C}—O—R \quad + \quad H_2O$$

$$R—\overset{\overset{\displaystyle OH}{|}}{CH}—R \quad \underset{\text{reduction}}{\overset{\text{oxidation}}{\rightleftharpoons}} \quad R—\overset{\overset{\displaystyle O}{\|}}{C}—R$$

$$—\overset{|}{C}=\overset{|}{C}— \quad \underset{\text{dehydrogenation}}{\overset{\text{hydrogenation}}{\rightleftharpoons}} \quad —\overset{\overset{\displaystyle H}{|}}{C}—\overset{\overset{\displaystyle H}{|}}{C}—$$

$$R—\overset{\overset{\displaystyle O}{\|}}{C}—Y \quad + \quad X^- \quad \overset{\text{substitution}}{\rightleftharpoons} \quad R—\overset{\overset{\displaystyle O}{\|}}{C}—X \quad + \quad Y^-$$

$$—\overset{\overset{\displaystyle OH}{|}}{C}—\overset{\overset{\displaystyle H}{|}}{C}— \quad \underset{\substack{\text{hydration}\\\text{(addition)}}}{\overset{\substack{\text{dehydration}\\\text{(elimination)}}}{\rightleftharpoons}} \quad —\overset{|}{C}=\overset{|}{C}— \quad + \quad H_2O$$

Chapter 1

....

Chemistry, energy, and metabolism

Chapter summary

This chapter describes three main topics: the effects of energy considerations on biochemical reactions; the role of enzymes as catalysts; and the types of chemical interactions that occur between biological molecules. A number of review topics are presented. These include the application of Gibbs free energy equations and the calculation of the equilibrium constant to measure the role of energy considerations in determining the direction in which chemical reactions proceed. In addition, the application of the principle that ΔG values can be added when reactions are coupled through a common intermediate is reviewed. An understanding of the effect of pH on the ionization of acids and bases is needed, and an appendix is included with the chapter to aid in the student's review of this material. The sections covering the basics of enzyme catalysis and the importance of ATP as an energy carrier offer an introduction to some basic biochemical principles, which will be expanded upon later in the text. An introduction to the class of attractive forces known as 'weak bonds' includes the ionic bonds, the hydrogen bonds, and the van der Waals forces. This lays important groundwork for a deeper understanding of the role that these forces play in the way in which biological molecules recognize each other, an idea that will be built upon throughout the rest of the text.

Learning objectives

❑ The relationship between the Gibbs free energy (ΔG) and the direction in which a chemical reaction will spontaneously go.

❑ The relationship between ΔG and the enthalpy (ΔH) and entropy (ΔS) changes of the reaction.

❑ The mathematical relationship between $\Delta G^{0\prime}$ and the equilibrium constant.

❑ The mathematical form of the equilibrium constant in terms of the concentrations of the products and the reactants at equilibrium.

❑ The role of essentially irreversible reactions in the overall flow of material through metabolic pathways.

❑ The function of enzymes as catalysts.

❑ The basic form of the Michaelis–Menten equation.

❑ The difference between catabolic and anabolic reactions.

❑ The role of the Sun in forming high-energy food molecules.

❑ The role of pH in determining the ionization state of a molecule given its pK_a value.

❑ The way in which a weak acid and its conjugate salt can form a buffer.

❑ The role of high-energy phosphate in the structures of adenosine and its phosphorylated derivatives.

❑ The principle of adding reactions with a common intermediate to couple reactions that are endergonic with reactions that are exergonic producing a net exergonic reaction.

❑ The importance of the class of intermolecular forces known as the weak bonds, including ionic bonds, hydrogen bonds, and van der Waals forces.

A walk through the chapter

Energy changes associated with chemical reactions

Energy considerations are the bottom line in all chemical transformations. Just like water flowing downhill, chemical materials combine and are changed in accord with the flow of energy from a high potential state to a lower one. Metabolism is the biological equivalent of a power plant, in many ways similar to an internal combustion engine. Energy is required for most of the everyday things biochemical systems do. They meet this energy need through the metabolism of food, which enters the system with a high 'chemical potential' and leaves as waste with a lower 'chemical potential'. Every chemical reaction can be assessed according to its accompanying change in energy, expressed mathematically as the Gibbs free energy (ΔG), a composite of two other fundamental energy parameters, enthalpy (ΔH) and entropy (ΔS). The direction of the energy change (into or out of the system) is designated by a positive or negative sign, respectively, placed in front of the value of ΔG. Changes in this energy parameter are calculated by taking the energy of the final state minus the energy of the initial state. Thus a decrease in the free energy of the system, with the final value being less than the initial value, would result in a negative value for ΔG.

ΔG is a measure in energy units of how far (and in which direction) the system is from equilibrium. A large negative value tells us that the reaction will proceed from left to right (as written). A positive ΔG tells us that the reaction will go in the opposite direction and a zero value for ΔG tells us that the system is at equilibrium and that there will be no net reaction. Biochemical ΔG values are tabulated under standard conditions (25°C, pH 7.0) and at equal concentrations of all starting materials and products (1.0 M). These values are designated $\Delta G^{0\prime}$.

Complete the following

Direction of energy flow	Sign of ΔG	State with the higher amount of free energy	Direction in which reaction will spontaneously go
Into the system	Positive	............................(1)	Right to left toward reactants
........................(2)	Negative	Reactants	Left to right toward products
No net flow of energy	............(3)	Equal in energy	................................(4)

Answers: (1) Products; (2) Out to the system; (3) Zero; (4) No change.

$\Delta G^{0\prime}$ is related to the equilibrium constant (the ratio of products to reactants at equilibrium) by a logarithmic relationship. By using this mathematical equation and the values of the temperature (in Kelvin, °C + 273) and the gas constant in energy units ($R = 8.3$ J mol^{-1} K^{-1}), one can calculate back and forth between the equilibrium constant K and the value of the free energy difference between the products and reactants under standard conditions. Complete the following equations. For the reaction A + B $\Leftrightarrow$ C + D,

$K_{eq}{}' = $ ....   and

$\Delta G^{0\prime} = $ ....

For a more complete understanding of a particular equilibrium situation not at standard conditions (cf. inside the cell), the value of $\Delta G'$ (without a superscript '0'), which is concentration-dependent, must be calculated. Fortunately, the magnitude of the effect of concentrations on the value of $\Delta G'$ is limited so that only those reactions with small values of $\Delta G^{0\prime}$ (close to zero) need be considered sufficiently sensitive to cellular concentrations to be reversible under physiological conditions. This is because reactions with either very large (+) or very large (−) values for $\Delta G'$ will be driven either very far to the left or very far to the right, respectively. The important consequence of this is that, since most cellular chemical transformations occur in a series of individual steps known as pathways, the irreversible steps (those with a large negative $\Delta G'$) act as one-way control valves for the flow of material through the pathway. Because of this, the entire metabolic pathway is not strictly reversible and the organism can thus control synthetic and degradative steps in a coordinated fashion by specifically regulating the flow of material at these irreversible steps.

A note on the use of $\Delta G'$ (without the superscript '0'). A general form of the Gibbs equation is: $\Delta G' = \Delta G^{0'} + RT \ln([\text{products}]^P/[\text{reactants}]^R)$, where the notation indicates that the concentration of products raised to the power of the coefficient for each product is multiplied together and then divided by a similar product of each of the reactant concentrations. This G' is then not a constant but a measure of how far the system is from equilibrium when the actual concentrations of products and reactants are plugged into this expression. $\Delta G^{0'}$, which is a constant, is strictly the energy difference between the system at equilibrium and the system at standard conditions, with all concentrations equal to 1 M. However, $\Delta G^{0'}$ can be used for comparison when comparing two reactions under the same set of conditions. Under these conditions the relative differences in the $\Delta G^{0'}$ values will be similar enough for most purposes whether the conditions for comparison are at standard state or at the actual concentrations found inside the cell.

Enzymes control the direction and speed of chemical reactions

Predicting the direction of a chemical reaction by the amount of energy associated with a given chemical transformation ($\Delta G^{0'}$) does not tell us about the speed or rate (change per time) of the reaction. This is where enzymes come in. Enzymes are a class of biomolecules. They are catalysts with a very specific job. Most enzymes are very big protein molecules and are composed of amino acids. One very famous generalized chemical equation often used to describe enzyme-catalysed reactions is the Michaelis–Menten equation in which the enzyme, substrate, transition state, and product each has its own symbol, (E, S, $S^{\ddagger}$, and P, respectively). In order for the substrate (reactant) to be changed into the product an energy barrier (hump) must be crossed. The higher the hump (activation energy), the slower the reaction will go. Enzymes speed up reactions by lowering the activation energy. Complete the following showing that the size of the hump is less when the reaction is catalysed by an enzyme.

Complete the following

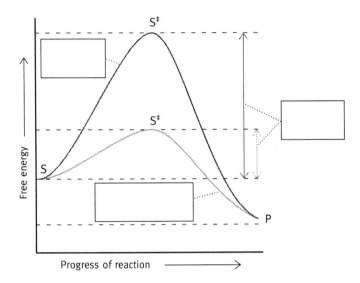

Notice that by following the catalysed pathway the substrates (S) can be changed into products (P) faster. Two models developed to explain how enzymes do this are the lock and key model and the induced fit model. Both are also useful for explaining why each enzyme catalyses only a certain reaction. The degree of acceleration an enzyme causes is sensitive to many parameters, such as pH and temperature and the presence of cofactors and inhibitors.

Chemical structures help explain how things happen

Chemical reactions that have overall favourable changes in free energy ($\Delta G' < 0$) occur 'energetically' unassisted, that is, with the release of free energy. The term used to describe such a reaction is 'spontaneous' and this term has a very specific meaning in this context. It does not imply that the reaction occurs 'quickly', but rather that it is favourable, occurring from left to right as written. The breaking down of 'food' molecules into lower energy molecules, which is known as catabolism, and the synthesis of other useful molecules with a high chemical potential from simpler precursors, which is known as anabolism, are processes that are coupled together. The Sun provides the energy for the formation of high-energy food molecules from simple precursor molecules. There are a number of figures in the main text that are related to one another in that they diagram this energy relationship between chemicals that are low on the free energy scale (such as CO_2 and H_2O) and those that are high on this scale such as 'food molecules'. Light energy is used to convert the low-energy $CO_2 + H_2O$ into food molecules by a process known as photosynthesis; this is then coupled to the reverse catabolic reactions, which require

Complete the following

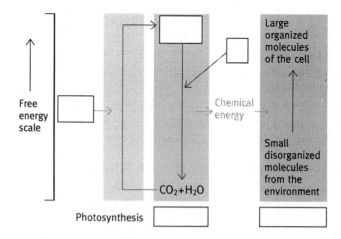

oxygen and release the stored energy from the Sun. The energy released form these catabolic steps with $\Delta G < 0$ is then coupled to anabolic chemical reactions, which perform the necessary processes within the cell. The role of high-energy phosphate compounds is to couple the energy obtained from the catabolism of food stuff to that used for doing biochemically useful work.

Biochemical systems often 'couple' an endergonic reaction to an exergonic one using a common energetic intermediate. In many cases the nature of the intermediate involves a high-energy phosphate compound. Phosphate in its low-energy form is called inorganic phosphate and is derived from the parent molecule phosphoric acid, H_3PO_4. Each of the four oxygens of this molecule is bonded to the central phosphate. One of them has a double and the other three have single bonds to the phosphorus. Each of the oxygens that are singly bonded to the phosphorus also has a single bond to a hydrogen. These hydrogen–oxygen bonds can be broken in what is called an acid dissociation reaction. The hydrogen departs as a proton (H^+) leaving behind its electron, which results in a negatively charged oxygen. Since this can happen three times for phosphoric acid, there are four possible different structures for inorganic phosphate having charges of 0, −1, −2, and −3. The proportional amounts of these species present in solution are pH-dependent. There are three pK_a values associated with inorganic phosphate (H_3PO_4) since it has three acidic protons. Note that the pK_a is a number that reflects the acid strength of the proton. The first of the protons to be lost from H_3PO_4 is very acidic, $pK_{a_1} = 2.2$, while the second is only moderately so, $pK_{a_2} = 7.2$, and the third proton with a $pK_{a_3} = 12.3$ is not very acidic at all. An easy way to work with these numbers is to remember that, when the pH equals the pK_a, there are an equal amount (50% of each) of the two chemical forms involved in the acid dissociation. The chemical form *with* the proton is referred to as the acid and the one that *has lost* the proton is referred to as the salt or conjugate base. For more details see the appendix at the end of Chapter 1 in the main text.

The appendix on buffers and pK_a values should be read carefully. Because so much of biochemistry occurs in water, the effects of the pH and the acid/base properties of biomolecules are very important. Since pH is defined in terms of the negative log of the hydronium ion concentration [H$^+$] and the buffer effect of weak acids and bases is related to their equilibrium constant (K_a), a most convenient equation to describe this connection is the Henderson–Hasselbalch equation

$$pH = pK_a + \log ([salt]/[acid])$$

where the acid is the species that can donate the proton and the salt is the chemical form of the same compound which has already lost the proton. Thus, the ratio of the protonated and unprotonated forms is dependent on both the pH and the strength of the acid. (In this equation the acid dissociation equilibrium constant K_a is expressed as the pK_a which is the negative log of this number.) Keeping in mind that the log of a number that is >1 is positive and that the log of a number that is <1 is negative (with log (1) = 0), one can see that, if the pH > pK_a, the salt (unprotonated) form will dominate and, if the pH < pK_a, then the acid (protonated form) will dominate; if the pH =pK_a, then there will be equal (50/50) amounts of each. Note that in this application all acid/base equilibria are viewed from the point of view of an acid dissociation, such that the pK_a of a protonated base such as $R - NH_3^+$ is around 9.2 which makes using an expression for the K_b of the R—NH$_2$ form unnecessary.

High-energy phosphate compounds are made by replacing the weak hydrogen–oxygen bond in inorganic phosphate with a bond from this oxygen to a heteroatom (an atom other than oxygen), typically a carbon, nitrogen, or, quite commonly, another phosphorus atom. Compounds of this sort can be either esters or anhydrides. (See the organic chemistry review in this companion.) Depending on which heteroatom is linked, compounds of different energies can be made. In addition, this process can occur on more than one of the oxygens of inorganic phosphate, producing, for example, pyrophosphate and the adenosine phosphates, a family of molecules that contain 1, 2, or 3 phosphates linked by high-energy bonds (AMP, ADP, and ATP). Identify the parts of each of the following molecules.

Complete the following

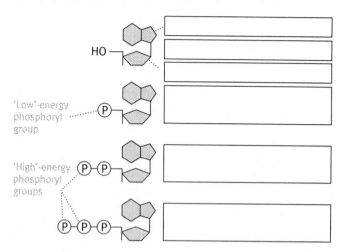

The energy stored in these compounds can be released through a reaction involving H_2O called hydrolysis. This is the exergonic reaction, which is coupled to and thus drives the endergonic biochemical reactions needed by the organism. High-energy phosphates are often abbreviated with a circled phosphorus and the low-energy inorganic phosphate is written as P_i. Cycling of the phosphorus between the high-energy and low-energy forms is one of the links between catabolic and anabolic reactions.

One common mechanism for coupling the hydrolysis of high-energy phosphate compounds with endergonic reactions makes use of the principle of additivity among reactions with a common intermediate, where the common intermediate is a chemical species that is the product of one step and the reactant of another. When this is done, the intermediate does not appear in the overall or net chemical reaction. A simple example might be one in which the reactions A → B and B → C can be added to give A → C. If the first reaction A → B releases more energy than the second one B → C requires, the net reaction A → C will be favourable energetically. Energetically favourable means that $\Delta G' < 0$, and that the equilibrium will lie to the right, that is, compound A will 'spontaneously' react to give C.

Two examples are shown in the text. In the first, two reactions are added and X–P is the intermediate.

Reaction 1: $X\!-\!OH + ATP \Leftrightarrow X\!-\!P + ADP$

Reaction 2: $X\!-\!P + Y\!-\!H \Leftrightarrow X\!-\!Y + P_i$

Sum of 1 and 2: $X\!-\!OH + ATP + Y\!-\!H \Leftrightarrow X\!-\!Y + ADP + P_i$

The other example shows how three reactions can be added. The occurrence of X–AMP couples the first and second reactions and the intermediate PP_i (also called pyrophosphate) couples the first and third reactions

Reaction 1:	X—OH + ATP $\Leftrightarrow$ X—AMP + PP_i
Reaction 2:	X—AMP + Y—H $\Leftrightarrow$ X—Y + AMP
Reaction 3:	PP_i + H_2O $\Leftrightarrow$ $2P_i$
Sum:	X—OH + ATP + Y—H + H_2O $\Leftrightarrow$ X—Y + AMP + $2P_i$

Note that neither of the intermediates appear in the final (summation) equation. When ATP is used to form AMP and pyrophosphate, the AMP must be cycled back to ATP through ADP with the use of a kinase. As is often the case, pyrophosphate (PP_i) is hydrolysed to $2P_i$ releasing even more energy to 'drive' the net reaction.

A complete understanding of the roles that chemical structures play in how things happen must go beyond covalent bonds to include interactions of a more subtle nature. A class of interactions known as weak bonds (or forces) are mainly responsible for the types of chemical interactions that occur between molecules when covalent bonds are not being broken and formed. The main text focuses on three types: ionic bonds; hydrogen bonds; and van der Waals attractions. Ionic interactions are based simply on the attractive forces between oppositely charged (positive and negative) groups. This type of force is always present when one is dealing with charged ionic species. The hydrogen bond is a special extension of the dipole–dipole interaction, which occurs between molecules with an unsymmetric distribution of electrons. A hydrogen bond almost looks like a hydrogen bridge, with two electronegative atoms sharing the same hydrogen. There is an optimum length and angle to the hydrogen bond. Van der Waals attractions are the weakest of the weak bonds but here as with the other weak bonds a great deal of stabilizing energy can be gained by the additive effect of a large number of small interactions. The cumulative effect of many weak bonds can often explain some of the macroscopic physical properties we often observe, such as whether a solid will dissolve in a given liquid (solubility) or whether two liquids will mix and be miscible or form two immiscible layers. The occurrence of these weak bonds is an important theme of this book and will be very important in deepening our understanding of how the interactions of biomolecules make the biochemical system work.

Review of problems from the end of Chapter 1

- Since 50% of the energy intake is used to form ATP, and it requires 55 kJ for each mole of ATP to be formed, the man can make approximately 91 moles, or 50 141 grams of ATP. The number seems ridiculous, but is reasonable since the ATP is not stored, but used as quickly as it is made.

- $\Delta G^{0\prime}$ numbers are useful for comparison, but are specific for standard conditions. When doing calculations at conditions other than standard state ([ ] = 1 M), a different value, $\Delta G'$ (without the superscript '0'), must be used.
- The difference in energy between the hydrolysis of AMP and that of ATP and ADP results from the fact that the bonds connecting the terminal phosphate to the rest of the molecule are different. In the case of ATP it is an anhydride while in the case of AMP it is a simple ester.
- Increasing temperature generally increases the rate of a reaction, but enzymes are not always stable at higher temperatures.
- Water is polar and 'like' polarities dissolve in 'like' solvents. H-bonds are also important.
- Although some enzymes use the higher-energy phosphate bond energies of both phosphates at once, AMP is recharged one phosphate at a time.
- One can add reactions and ΔG values. In the cell the following three reactions can occur.

$$X—OH + Y—H \Leftrightarrow X—Y + H_2O$$
$$ATP \Leftrightarrow AMP + PP_i$$
$$PP_i \Leftrightarrow 2P_i$$

$$\Delta G = (+10 \text{ kJ mol}^{-1}) + (-32.2 \text{ kJ mol}^{-1}) + (-33.4 \text{ kJ mol}^{-1}) = -55.6 \text{ kJ mol}^{-1}$$

In an *in vitro* system such as a test tube containing only the first enzyme, the reaction catalysed by the pyrophophatase ($PP_i \Leftrightarrow 2P_i$) does not occur, and the ΔG value is less exergonic.

- Make a table to help remember the following facts.

Types of weak bonds	Energy (kJ mol^{-1})
Ionic bonds	20
H-bonds	12–29
van der Waals	4–8

The individual interactions are weak, and form readily, but play a significant role in determining biochemical specificity because they are additive.

- Each acidic proton has a pK_a. The value of the pH determines at what point each proton is removed. The pK_a is the pH at which one-half of the molecules have this proton removed, and this is the point at which the compound can act as a buffer, because addition to or removal of protons from the system at this point will change the proportion of buffer molecules that are protonated rather than changing the pH of the solution. The magnitude of this effect is limited to the total concentration of the buffering molecules.

- Glucose is thermodynamically unstable with respect to the products CO_2 and water, but it is kinetically stable toward this reaction. This means that energy will be released as the reaction occurs, but that the reaction will go at a rate too slow to be observed. ΔG depends only on the energy difference between the starting point and the ending point and thus indicates the direction in which the reaction has a tendency to occur. However, the value of ΔG tells us nothing about the rate. Another interesting example of this is the fact that the diamond form of carbon is more stable than the graphite form. What a pity this favourable reaction is so slow! Biochemistry students often get the idea that, since all chemical reactions have a tendency to move toward equilibrium, this is the normal state of affairs. Nothing could be further from the truth. Biological systems strive to obtain a steady-state condition as far from equilibrium as possible. Equilibrium for a biological system is equivalent to cell death since all chemical reactions would stop and there would be no net energy change.

- Enzymes are not magic. The same energy considerations that affect normal chemical reactions always apply to the same reaction catalysed by an enzyme. The increase in rate seen in the enzyme-catalysed reaction comes from the unique interactions that can occur between the substrate and the complex enzyme molecule. This is seen energetically as a decrease in the activation energy hump.

Additional questions for Chapter 1

1. List two examples of chemical reactions that release energy as they proceed.
2. Compose your own definition of the following terms:

    (a) enthalpy; (b) entropy; (c) free energy.
3. Show that the value of $\Delta G^{0\prime}$ is equal to zero when the value of the equilibrium constant K is equal to 1.0. What does this unique situation imply about the energy differences between the products and the reactants?
4. Calculate the value for the equilibrium constant K when $\Delta G^{0\prime}$ has a value of -11.4 kJ mol^{-1} at 310 Kelvin. What does this calculation say about the relative amounts of the products and reactants at equilibrium in this system?
5. Explain in your own words the differences and similarities between the two theories used to explain how enzymes function, that is, the lock and key model and the induced fit model.
6. What is the structure of the major species of the phosphate ion at pH = 7.
7. What is the $[H^+]$ at pH = 10?
8. What is the pH when the $[H^+] = 0.025$ M?
9. Which is more acidic, the imidazole group of histidine ($pK_a = 6$) or the carboxylic group on formic acid $K_a = 1.74 \times 10^{-5}$?

10. Draw a titration curve showing the buffering region for adding HCl to an already deprotonated (pH = 10) solution of acetic acid (pK_a = 4.76).

11. What is an appropriate pH range in which the compound ammonia (NH_3, pK_a = 9) can act as a buffer?

12. The pK_a of lactic acid is 3.1 at 25°C. At what pH value will the ratio of the protonated to unprotonated form of lactic acid be 4 to 1?

13. Explain the concept of a buffer in your own words.

14. Which weak forces would be operating in the process by which a salt crystal such as NaCl would dissolve in water?

15. List the types of H-bonds that could form between the following molecule and water.

16. Explain, using terms you have just learned in this chapter related to the weak forces, why water and oil don't mix.

Chapter 2

....

The structure of proteins

Chapter summary

This chapter describes the structures and chemical properties of the amino acids and the principles used to distinguish between the different levels of protein structure. Some examples of membrane proteins and conjugated proteins are presented. The connection between molecular structure and physiological function is presented with examples taken from soluble enzymes and proteins that occur in connective tissues.

Learning objectives

- ❑ Structures of the 20 amino acids.

- ❑ Distinctions between globular and fibrous proteins.

- ❑ Classifications of amino acids based on R-group polarities.

- ❑ The four levels of protein structure and the nature of the attractive forces involved at each level.

- ❑ The three main types of secondary structure.

- ❑ The role of disulfide bonds and the oxidation/reduction reactions that these linkages undergo.

- ❑ The connection between protein structure and function.

- ❑ Examples of the common types of conjugated proteins.

- ❑ The importance of correct protein folding.

A walk through the chapter

Amino acid chemistry

Proteins are polymers of ....(1).   Both the amine nitrogen and the ....(2) part of an amino acid can be ionized to form a doubly charged zwitterion. Each amino acid, except ....(3), has a ....(4) carbon with four different groups. Each of these has the L-stereoconfiguration. Amino acids are a group of organic molecules with similar backbones and 20 different R-groups. It is the chemical nature of the R-group that determines the properties of the amino acid within the protein polymer. The ....(5) of the amino acids can be grouped into two broad classes, ...(6) and ....(7).   In accord with the principle of like dissolves like, polar groupings are ....(8)   and ....(9)   groups are fat- (or oil-) soluble. There are 10 nonpolar amino acids, which can be further grouped into the aliphatic and ....(10) amino acids. There are 10 polar amino acids, which can be grouped into those that are positive, ...(11), and just simply polar. The charges on the positive and negative amino acid side chains, along with the charges on the N and C termini are due to acid/base reactions and so are pH-dependent. Biochemists are mostly interested in the state of these chemical groups at pH = 7, but a more general approach to this reasoning tells us that, if the pH > pK_a then the group is deprotonated (ionized (–) for a carboxylic acid), while if the ....(12)   then the group is protonated (ionized (+) for an amine). When the pH = pK_a, one has equal amounts of the two forms.

Answers: (1) amino acids; (2) carboxylic acid; (3) glycine; (4) chiral; (5) R-groups; (6) polar; (7) nonpolar; (8) water-soluble; (9) nonpolar; (10) aromatic; (11) negative; (12) pH < pK_a.

The structures of the 20 common amino acids

Nonpolar aliphatic

$^{+}H_3N-CH-COO^{-}$
|
H
glycine (Gly)

$^{+}H_3N-CH-COO^{-}$
|
CH_3
alanine (Ala)

$^{+}H_3N-CH-COO^{-}$
|
CH
|
CH_3 CH_3
valine (Val)

$^{+}H_3N-CH-COO^{-}$
|
CH_2
|
CH
|
CH_3 CH_3
leucine (Leu)

$^{+}H_3N-CH-COO^{-}$
|
$CHCH_3$
|
CH_2
|
CH_3
isoleuine (Ile)

$^{+}H_2N-\underset{|}{C}-COO^{-}$
H
proline (Pro)

Basic

$^{+}H_3N-CH-COO^{-}$
|
CH_2
histidine (His)

$^{+}H_3N-CH-COO^{-}$
|
$(CH_2)_4$
|
NH_3^{+}
lysine (Lys)

$^{+}H_3N-CH-COO^{-}$
|
$(CH_2)_3$
|
NH
|
$HC=NH_2^{+}$
|
NH_2
arginine (Arg)

Aromatic

$^{+}H_3N-CH-COO^{-}$
|
CH_2
phenylalanine (Phe)

$^{+}H_3N-CH-COO^{-}$
|
CH_2
|
OH
tyrosine (Tyr)

$^{+}H_3N-CH-COO^{-}$
|
CH_2
|
NH
tryptophan (Trp)

Alcohols

$^{+}H_3N-CH-COO^{-}$
|
CH_2
|
OH
serine (Ser)

$^{+}H_3N-CH-COO^{-}$
|
CH—OH
|
CH_3
threonine (Thr)

Amides

$^{+}H_3N-CH-COO^{-}$
|
CH_2
|
C=O
|
NH_2
asparagine (Asn)

$^{+}H_3N-CH-COO^{-}$
|
CH_2
|
CH_2
|
C=O
|
NH_2
glutamine (Gln)

Acidic

$^{+}H_3N-CH-COO^{-}$
|
CH_2
|
COO^{-}
aspartate (aspartic acid) (Asp)

$^{+}H_3N-CH-COO^{-}$
|
CH_2
|
CH_2
|
COO^{-}
glutamate (glutamic acid) (Glu)

Containing sulfur

$^{+}H_3N-CH-COO^{-}$
|
CH_2
|
SH
cysteine (Cys)

$^{+}H_3N-CH-COO^{-}$
|
CH_2
|
CH_2
|
S
|
CH_3
methionine (Met)

Complete the following

Classification based on R-group	Amino acids
Aliphatic	....................................(1)
....................................(2)	Phe, Tyr, Trp
Sulfur-containing	....................................(3)
....................................(4)	Ser, Thr
Basic	....................................(5)
Acidic	....................................(6)
....................................(7)	Asn, Gln

Answers: (1) Gly, Ala, Val, Leu, Ile, Pro; (2) Aromatic; (3) Cys, Met; (4) Alcohols; (5) His, Lys, Arg; (6) Asp, Glu; (7) Amides.

Levels of protein structure

The shape of a protein (native configuration) is related to its function. The types and positions of the amino acids that make up the protein polymer determine this shape. Proteins can be grouped into two classes based on overall shape: these are the ....(1) proteins and the ....(2) proteins. Globular proteins, particularly those that are water-soluble, fold into an active conformation that arranges the polar water-liking ....(3) groups on the outside and the nonpolar water-...............(4), hydrophobic groups on the ....(5). A convenient way to discuss and compare the levels of complexity contained within a protein's structure is in terms of the primary, secondary, tertiary, and quaternary structure. The ....(6) level of structure includes the sequence of amino acids that makes up the protein backbone. The ....(7) of one amino acid is linked with the ....(8) of a second to form an amide bond. The amino group of the first and the carboxylic group of the last are unbonded and ionized. By convention the primary sequence is numbered from *N*-terminus to *C*-terminus. The secondary level of structure involves the arrangement of the polypeptide backbone. There are three common types of secondary structure: the ....(9);   the   ....(10)   sheet;   and   the ....(11). The α helix and β-pleated sheet structures are stabilized by intramolecular (within) ....(12). Not all orientations of the backbone are possible since the amide bond has 40% ....(13) character as a result of electronic resonance. Most helices are right-handed; the β sheets are more extended than the α helix and can be formed between chains running in a ....(14) or antiparallel direction. The tertiary level of structure includes the positions of all of the R-groups and, if the peptide has only a single amino acid chain, this level involves the complete native three-dimensional arrangement of all the atoms.

Answers: (1) globular; (2) fibrous; (3) hydrophilic; (4) fearing; (5) inside; (6) primary; (7) carboxylic acid; (8) amino group; (9) α helix; (10) β-pleated; (11) random coil; (12) hydrogen bonds; (13) double bond; (14) parallel.

Additional structural principles

Much of the energy stabilization responsible for keeping the protein in its proper orientation comes from the numerous weak forces that occur between neighbouring groups. In addition, covalent S—S disulfide bonds through the amino acid ....(1) are quite common and greatly enhance the stability of a given structure. *Take time to notice the figures* and try to identify the different levels of protein structure in each. The quaternary level of protein structure is only appropriate when ....(2) polypeptide chain is involved. These are known as ....(3) or multisubunit proteins. Proteins can also be ....(4) or linked with other chemical groups such as sugars. In this case the resultant glycoproteins can play special roles on the exterior of cells. Modern biochemists are very interested in the concept of protein ....(5). A domain is a region that is thought to fold and often function separately from the rest of the protein molecules. Evidence for this can be found in proteins that perform similar functions on different substrates where the substrate-binding domains are different, while the catalytic site of action are similar. The fibrous class of proteins play specialized ....(6) roles outside of the cell. It is fascinating how structural differences on the molecular level can have profound effects on the macroscopic properties, with differences ranging from the strength and rigidity of a ....(7) to the flexibility and elasticity of skin and ....(8) tissue. One of the many differences between fibrous proteins and globular proteins is that the globular proteins tend to be ....(9), and the fibrous proteins tend to form ....(10). Many types of covalent and noncovalent forces give these fibrous proteins their unique properties. The mechanism (how it happens) of protein folding is not yet completely understood and as such may be one of the next frontiers.

Answers: (1) cysteine; (2) more than one; (3) oligomeric; (4) conjugated; (5) domains; (6) structural; (7) tendon; (8) lung; (9) water-soluble; (10) solid structures.

Review of problems from the end of Chapter 2

- The sequence of amino acids is read from left to right and constitutes the primary structure of a protein.
- Denatured proteins often form precipitates; one example is the curdling of milk.
- Amino acids are classified according to the structure and properties of their R-group.
- The pK_a values of amino acids can be grouped according to the nature of each of the chemical functionalities present; for example, all carboxylic acid groups have a pK_a around 2.
- In addition to the N-terminal and C-terminal amino acids, certain R-groups can contribute to the charge on a polypeptide.
- The classification of protein structure into levels is useful for distinguishing the types of stabilizing interaction involved and the complexity of the overall structure.
- The level of secondary structure can be divided into three main categories.
- Proteins as a group of biomolecules are extremely versatile in function.
- For many polymers the characteristic property on the molecular level (in the case of elastin its stretchability) is often carried over into the properties of the substance as a whole.

Additional questions for Chapter 2

1. Polypeptides can be named by listing the individual amino acids (from N- to C-terminal) with the suffix -yl replacing the -ine (or -ate) for most amino acids and replacing the -e for asparagine, glutamine, and cysteine. The name of the last amino acid in the chain is left unchanged; thus alanylvaline is a dipeptide. Draw the structure of glycylhistidylphenylalanylproline at pH = 7.
2. How does secondary structure differ from tertiary structure?
3. Why is proline not found in the α-helix structure?
4. Which amino acid is involved in disulfide linkages?
5. What is the nature of the prosthetic group in a glycoprotein?
6. How do the structures of collagen and elastin relate to their biological functions?

Chapter 3

···

The cell membrane—a structure depending only on weak forces

Chapter summary

This chapter describes the functions of cell membranes and the chemical principles that govern their structures. The chemical structures of the major classes of lipid molecules are presented as well as the roles of carbohydrates and proteins, in specific examples related to membrane transport, signal transduction, cell shape, and cell–cell interactions.

Learning objectives

- ❑ The properties of polar, nonpolar, and amphipathic molecules.

- ❑ The difference between a lipid bilayer and a micelle.

- ❑ The general structure of fatty acids and how they combine to form triacylglycerols (TAG).

- ❑ The structures of the major glycerophospholipids (PE, PC, PI, and PS).

- ❑ The structural features that make sphingolipids different from glycerophospholipids.

- ❑ The different types of movement that individual lipid molecules can undergo.

- ❑ The difference between *cis* and *trans* double bonds and their effect on membrane fluidity.

- ❑ The types of molecules that can and cannot readily cross lipid membranes.

- ❑ A general understanding of the role of proteins in the membrane.

- ❑ Four physiological functions of membranes and specific examples of each.

- ❑ Six major classes of membrane structures within eukaryotic cells and their uniqueness.

A walk through the chapter

The structure of cell membranes

Cell membranes function to hold together the contents of the cell. The lipid mole-cules that make up the membrane have the property of being ....(1), which means that each molecule contains a polar (hydrophilic; water-soluble) part and a nonpolar (.......................(2); water-insoluble) part. The individual amphi-pathic molecules arrange themselves so that the 'like dissolves like' principle applies; thus, the ....(3) groups stay in contact with the water, while the nonpolar tail groups stay away from the water and congregate together. The energe-tics of this arrangement involves only weak forces, and two general arrangements are common: a bilayer in which ....(4) rows of amphipathic molecules form an aqueous cavity (cf. a cell) and a ....(5), in which a single layer of amphipathic molecules form a droplet with a ....(6) core. Both types are important in biochemistry, but perform different functions. The structural feature of the polar lipids that determines whether a bilayer or a micelle is formed is the number of ....(7) molecules attached. The polar lipids found in cell membranes that form bilayers have *two* tails, while polar lipids with only ....(8) fatty acid tail form micelles.

Answers: (1) amphipathic; (2) hydrophobic; (3) polar head; (4) two; (5) micelle; (6) nonpolar; (7) fatty acid; (8) one.

Chemical structures of lipids

The term fat is applied in general to lipid molecules as a class because of their preference for dissolving in nonpolar organic solvents rather than in ....(1). The building block of most lipid molecules is the long chain fatty acid which con-tains a nonpolar ....(2) and a carboxylic acid. The formation of an ....(3) involving the carboxylic acid is used to attach the fatty acid tail portion on many lipid molecules. One important ester of fatty acids is the ....(4)   ester (TAG). This molecule is a neutral fat (not amphi-pathic and not found in membranes), but makes up a substantial part of our diet.

The three-carbon alcohol, ...(5), is the backbone of many lipid mole-cules. In conjunction with a phosphate group, these are known as glycero-phospholipids. Phosphatidic acid has two fatty acids and a phosphate esterified to the remaining OH of glycerol. Various head groups can also be esterified to the phosphate yielding molecules named with the phosphatidyl- prefix. These then are the amphipathic molecules (polar lipids) that make up a substantial portion of the membrane. There are a number of common head groups attached to phosphatidic acid. These are: ethanolamine forming phosphatidylethanolamine (......(6)), choline forming ....(7)   (PC or lecithin), serine forming phosphatidylserine (PS), and ....(8) forming phosphatidylinositol (PI). Cardiolipin is another novel glycerol-based phospholipid. A closely related class of lipids that are not based on the glycerol backbone is that of the sphingolipids. The parent compound is called sphingosine. The molecules ceramide and sphingomyelin are derived from sphingosine. When sugars are attached to the phosphate of a sphingosine-based lipid, one can get ....(9) and ganglio-sides. *N*-acylglucosamine and ....(10) are important carbohydrate (sugar) groups found in gangliosides.

Answers: (1) water; (2) hydrocarbon chain; (3) ester bond; (4) triacylglycerol; (5) glycerol; (6) PE; (7) phosphatidylcholine; (8) inositol; (9) cerebrosides; (10) sialic acid.

Complete the following

Abbreviations	Name	Structure
Not commonly abbreviated	Glycerol	$CH_2\text{—}OH$ $CH\text{—}OH$ $CH_2\text{—}OH$
FA (general)	18-carbon saturated fatty acid	$CH_3 (CH_2)_{16} CO_2^-$
TAG	Triacylglycerol	?
PE	?	?
?	Phosphatidylcholine	?
Not commonly abbreviated	Ethanolamine	$HOCH_2CH_2NH_3^+$
PS	?	?
PI	Phosphatidylinositol	?

Membrane fluidity and permeability

The phospholipid composition of cell membranes varies greatly between cell types. Lipids interact with each other by weak forces and are thus able to move within the bilayer (laterally) at a rapid rate. The movement of phospholipids from one half of the bilayer to the other is known as ....(1) and is limited because of the unfavourable interactions that occur between the ....(2) group of the phospholipid and the interior of the membrane composed of nonpolar hydrocarbon tails. The fatty acid tail components almost always have an even number of ....(3) atoms, and can be saturated (without ....(4) bonds) or ....(5) (containing one or more carbon–carbon double bonds). When these double bonds exist, they are almost always in the ....(6) con-figuration, which produces a kink in the chain structure. This kink is an important contributor to the fluidity of the membrane since it does *not* allow close packing of the hydrocarbon chains, thus making the membrane ....(7) fluid (less rigid). Cholesterol is an important component of most membranes and is also involved in modulating membrane fluidity.

Cell membranes made from lipid bilayers are flexible and have a self-sealing potential. This becomes important in the processes of ....(8), ...(9), and ....(10). In addition, membranes are selectively permeable. Nonpolar fat molecules, as well as some small polar mole-cules such as water, can cross the membrane easily but large polar molecules and ionized groups cannot easily cross the membrane, due to the same reason, that phospholipids do not readily flip-flop.

Answers: (1) flip-flop; (2) polar head; (3) carbon; (4) double; (5) unsaturated; (6) *cis*; (7) more; (8) cell division; (9) endocytosis; (10) exocytosis.

Proteins and membrane functions

Proteins can be associated with membranes in an integral or ....(1) manner. Integral proteins have regions of hydrophobic ....(2) that can span across the nonpolar hydrocarbon layer. Peripheral proteins typically have a hydrocarbon lipid 'anchor' to keep them associated with the membrane. Membrane proteins often act as attachment sites for sugars, which can play a number of physio-logical roles on the ....(3) of the cell.

Four physiologically important functions in which membranes are involved are: ....(4); ....(5) transduction; ....(6); and ....(7). The transport of molecules across a membrane can be active or ....(8) depending on whether it requires ....(9). Passive processes that nevertheless require a ....(10) are called facilitated. Gated pores or channels are an example of facilitated passive transport: the protein controls the movement but the movement does not require energy. These can be controlled by a ....(11) or ....(12). Active transport requires ....(13) input for the transport to occur. The energy often comes from ....(14) hydrolysis, and is required because the direction of transport is opposite or against the existing gradient conditions between the inside and outside of the membrane. The Na^+/K^+ ATPase pump is a very well studied example of this type of transport. Na^+ is pumped ....(15) and ....(16) is pumped in at the expense of ATP hydrolysis, mediated through a protein conformational change. In this example, more than one molecule can be transported at the same time, and this is known as ....(17). If both molecules are transported in the same direction this is called ....(18), while if the directions are opposite it is known as ....(19). Signal transduction involves the chemical communication between cells. This necessarily involves membrane-mediated steps. Cells contain molecules that act as support structures, which are known collectively as the cytoskeleton. These molecules have special interactions with the proteins of a membrane to maintain cell shape. Cell–cell interactions are important in a variety of circumstances. 'Tight' junctions between cells keep the cells oriented properly and prevent leakage of molecules between the cells. Gap junctions constitute passageways between cells that are in contact and allow certain molecules to move between.

Answers: (1) peripheral; (2) amino acids; (3) outside; (4) transport; (5) signal; (6) cell shape; (7) cell–cell interactions; (8) passive; (9) energy; (10) protein; (11) voltage; (12) ligand; (13) energy; (14) ATP; (15) out; (16) K^+; (17) cotransport; (18) symport; (19) antiport.

Intracellular membranes

There are six major classes of membrane structures within eukaryotic cells. They are: the ....(1), which contains the DNA; the ....(2) reticulum, which has a rough and a smooth portion; the ....(3) apparatus, which sorts out synthesized proteins; the ....(4), which is the site of oxidative phosphorylation; the ....(5), which contain hydrolytic enzymes; and the ....(6), which metabolize certain molecules by oxidative reactions. Each of these has unique functions and structure.

Answers: (1) nuclear membrane; (2) endoplasmic; (3) Golgi; (4) mitochondrion; (5) lysosomes; (6) peroxisomes.

Complete the following

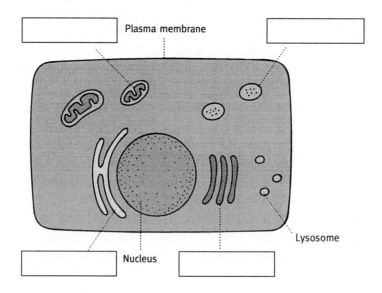

Review of problems from the end of Chapter 3

- There are easy ways to recognize the differences between a liposome and a micelle.
- The TAG molecule is not amphipathic.
- Proteins exit cells by a different mechanism than that used by simple ions.
- The word 'buffer' can be applied to more than just acid/base chemistry.
- There are many shorthand notations for drawing chemical structures.
- Phosphatidic acid + head group = glycerophospholipid.
- Make a list of the structures you will need to know. (Check with your instructor.)
- Lipid + sugar = glycolipid.
- The '*N*' designation in chemical names indicates where the acetyl group is attached.
- Closely packed fatty acids are rigid.
- Energy considerations govern cell processes.
- Facilitated transport is not an active process; it is passive.
- [Solute]$_{outside}$ $\Leftrightarrow$ [Solute]$_{inside}$ is a chemical reaction and has an equilibrium expression

  $$\Delta G = RT(2.303)\log_{10} \text{[Product]/[Reactant]}$$

  where *T* is in Kelvin, *R* is a constant, and the number 2.303 appears because of the use of log base 10.

- The intracellular balance of K^+ and Na^+ is very carefully controlled.

Additional questions for Chapter 3

1. Describe in words the structure of cardiolipin.
2. What unique kinds of linkages are present in ceramide?
3. What are some of the differences between the structure of a cerebroside and that of a ganglioside?
4. What is the effect of temperature on membrane fluidity and how might the composition of the fatty acids of an organism reflect the temperature of its environment?
5. If an organism is fed only two different types of fatty acids (assume that it can't make any on its own), how many different triacylglycerides can this organism make?
6. Describe in words the sequence of events that occurs when the Na^+/K^+ ATPase goes through one cycle.

Chapter 4

···

Digestion and absorption of food

Chapter summary

This chapter describes the metabolism of the three main classes of molecules used for food: proteins; carbohydrates; and fats. The four important regions of the body involved in food metabolism are also presented, as well as the chemistry that occurs as these food molecules are made ready for transport and use by the body.

Learning objectives

- ❏ The three main classes of food molecules.

- ❏ The four important regions of the body involved in food metabolism.

- ❏ The role of zymogens in the process of digestion.

- ❏ The mechanism for generating stomach acid and its effect on proteins.

- ❏ General reaction catalysed by a protease.

- ❏ The importance of stereochemistry as it relates to carbohydrates with an emphasis on glucose.

- ❏ The type of chemical bond used to link carbohydrates together.

- ❏ The mechanism of the transport of glucose through the membranes of the brush border cells.

- ❏ The role that emulsification plays in the metabolism of fat.

- ❏ The chemical reaction that describes the hydrolysis of TAG to yield free fatty acids.

- ❏ The chemical reaction that describes the esterification of free fatty acids to form TAG.

- ❏ The role of chylomicrons in the transport of lipids.

A walk through the chapter

The process of digestion

The three main classes of molecules that are considered to be food molecules are proteins, which are polymers of ....(1), carbohydrates (also called sugars or saccharides), which are often polymers of ....(2) units, and fats comprised mostly of ....(3).   The principal action of ....(4)   is to break down the polymers of ....(5)   into amino acids, the polymers of carbohydrates (polysaccharides) into monosaccharide units, and the esterified fats (mostly triacylglycerol) into free fatty acids. All three of these processes involve a similar chemical reaction, namely, ...(6).

There are four important anatomically distinct regions of the body where digestion takes place: the ....(7), the ....(8), and the ....(9) and ....(10).   Each plays a unique role in the digestive process. While the hydrolytic reactions of digestion are ....(11) and thus readily go to completion, the process of absorption in which material is moved across a membrane often requires the ....(12) of energy.

Answers: (1) amino acids; (2) monosaccharide; (3) neutral lipids; (4) digestion; (5) proteins; (6) hydrolysis; (7) mouth; (8) stomach; (9) small; (10) large intestines; (11) exergonic; (12) input.

The digestion of proteins

Zymogens (also known as proenzymes) are inactive forms of an enzyme that can become activated when needed. The utility of this relationship in the process of digestion is that enzymes that can catalyse the general breakdown of biomolecules can be safely synthesized and stored until needed without the problem associated with these molecules reacting with the components of the parent system. Zymogens are often named with the suffix 'ogen' or with the prefix 'pro' added to the name of the active enzyme (for example, pepsinogen is the zymogen of pepsin). The activation of a zymogen can often be an autocatalytic event involving the active enzyme itself causing the activation of more enzyme molecules. The mucus that surrounds the intestinal epithelial cells also affords protection in this regard.

Complete the following

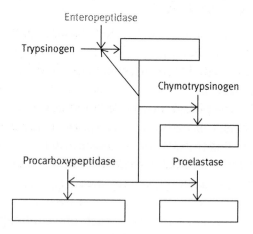

The digestion of proteins is initiated by denaturation (unfolding) in the presence of strongly acidic (HCl) conditions. To generate such acidic conditions a series of membrane transport and enzyme-catalysed reactions works to effect the generation of HCl in the stomach lumen.

Complete the following

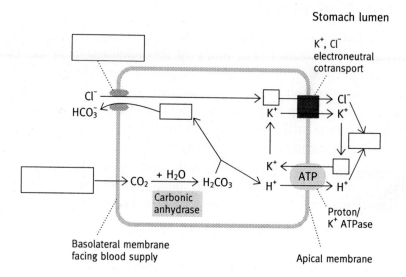

Notice how this cell has two distinct sides, the apical and the basolateral, and how the Cl^- and CO_2 are moved into the cell, combined with water so that HCl can be moved into the stomach lumen and HCO_3^- back into the blood supply.

Proteases

Enzymes that hydrolyse protein bonds are called ....(1). If they are specific for one of the peptide bonds within the molecule, they are called endopeptidases (or endoproteases): enzymes that cleave the *C*-teminal amino acid off of a peptide chain are called ....(2) and, similarly, ...(3) chop away at the amino end. The proteases responsible for digestion in the stomach operate well under ....(4) conditions; however, as the partially digested stomach contents (chyme) passes into the small intestine, the pH is raised and a new set of endopeptidases from the ....(5) continues the process of hydrolysing the polymer into monomer ....(6) units. The amino acids are absorbed from the intestine and transported eventually to the blood.

Answers: (1) proteases; (2) carboxypeptidases; (3) aminopeptidases; (4) acidic; (5) pancreas; (6) amino acid.

Carbohydrate digestion

Carbohydrates that are injested as monosaccharides can be ....(1) directly, while sugars linked together by glycoside bonds need to be hydrolysed before this can occur. Glucose is one of the most common monosaccharides and is a six ....(2) polyhydroxylated molecule, which readily forms a hexagonal ring containing a bridging ....(3). This ring can be formed with the hydroxyl at carbon atom 1 either in the ....(4), β position, or the ....(5), α position (with reference to the molecule drawn with carbon atom 6 in the up position). When sugars are linked through the ....(6) on carbon atom 1, the stereochemistry of this centre (α or β) is fixed. Individual sugars can also be linked through the oxygen atoms at other carbons on the ring, and chains of sugars can be branched by combinations of different linkages. A system of nomenclature has been developed to specify both the types of linkages and the stereochemistry involved. A common example is amylose which is a polymer of $\alpha(1 \rightarrow 4)$ linked ....(7) residues. Amylopectin contains ....(8) chains of glucose branched with $\alpha(1 \rightarrow 6)$ glycosidic bonds. The enzymes that hydrolyse sugars, like many enzymes, are given names that reflect what molecule (substrate) they act on followed by the suffix 'ase'. Thus you have amylase, ...(9), and sucrase, catalysing the hydrolysis of amylose, lactose, and ....(10), respectively. The process by which glucose is absorbed from the intestinal lumen involves the cotransport of ....(11) by the Na^+/K^+ ATPase.

Answers: (1) absorbed; (2) carbon; (3) oxygen; (4) up; (5) down; (6) oxygen; (7) glucose; (8) $\alpha(1 \rightarrow 4)$; (9) lactase; (10) sucrose; (11) Na^+ and K^+.

Complete the following

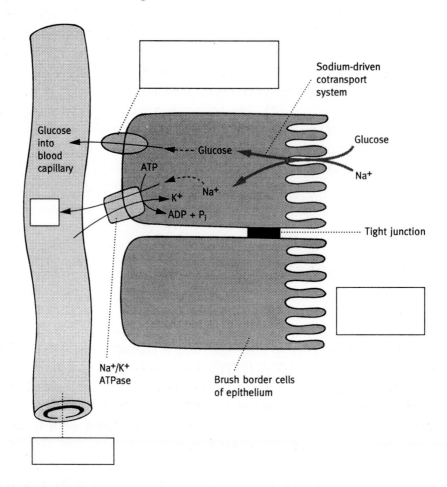

Fat digestion

The digestion and transport of fat presents a somewhat more involved process because the molecules of this class are not ....(1). Fats must be emulsified (broken down into very small droplets) before the lipase enzyme can hydrolyse the ....(2) in these lipids. Bile acids, a class of molecules structurally similar to ....(3) which are stored in the ....(4) and delivered through the duodenum, are responsible for this emulsification by acting as detergent-type molecules. Notice the amphipathic structure of cholic acid.

   Hydrolysis of ester bonds in neutral lipids produces ....(5) and the parent alcohols. Interestingly, once these molecules are transported into the cells lining the digestive system, they are re-esterfied into triacylglycerides and ....(6). For transport out of the intestinal cells, these lipids are repackaged into large spherical particles called ....(7), which belong to the class of biomolecules called lipoproteins. Chylomicrons eventually enter the bloodstream via the ....(8) system. The other classes of biomolecules that can function as food are treated, digested, and absorbed by pathways that are analogous (when possible) to those discussed above.

Answers: (1) water-soluble; (2) ester bonds; (3) cholesterol; (4) gall bladder; (5) free fatty acids; (6) cholesterol esters; (7) chylomicrons; (8) lymph.

Review of problems from the end of Chapter 4

- Many enzymes are produced as inactive zymogen forms as part of the system's mechanism for controlling their activity. Make a table to help remember these.

Active enzyme	Zymogen form
Pepsin	Pepsinogen
Chymotrypsin	Chymotrypsinogen
Trypsin	Trypsinogen
Elastase	Proelastase
Carboxypeptidase	Procarboxypeptidase

- Not all enzymes are produced as zymogens, and there are in fact additional protective measures needed when the proenzymes are activated.
- Proteases are activated by a number of mechanisms, which are often unique for the particular protease, but often involve autocatalytic enzymatic steps.
- The pancreas is susceptible to a number of risks due to the presence of the potentially powerful and indiscriminate digestive enzymes.
- Realizing that it is the monomer forms of food that are absorbed, make a table of the polymeric and monomeric forms of the three main classes of food molecules.

Food	Polymer	Monomer
Protein	Polypeptide	Amino acid
Sugar	Polysaccharide	Monomersaccharide
Fat	Membrane/storage droplet	TAG, phospholipid, etc.

- It would be a paradox for the system to get something for nothing. Clearly ATP is being hydrolysed and driving this process. The actual coupling of the endergonic and exergonic steps involves the Na^+/K^+ ATPase. As the Na^+ is pumped out into the blood capillary by this enzyme, the glucose is being cotransported in with Na^+ from the intestinal lumen.
- People suffering from lactose intolerance can often benefit by pretreating their dairy products so as to hydrolyse the lactose into glucose and galactose before it is consumed.

- There are a number of accepted shorthand notations for drawing chemical structures. A more complete structure of tripalmitate, a triacylglycerol (TAG) with three palimitic acids is shown below; other neutral fats include cholesterol and cholesterol esters.

$$
\begin{array}{l}
CH_2-\overset{\overset{\displaystyle O}{\|}}{C}-(CH_2)_{14}-CH_3 \\
\overset{|}{C}H-\overset{\overset{\displaystyle O}{\|}}{C}-(CH_2)_{14}-CH_3 \\
\overset{|}{C}H_2-\overset{\overset{\displaystyle O}{\|}}{C}-(CH_2)_{14}-CH_3
\end{array}
$$

- Emulsification increases the surface area of the fat exposed to the water. A somewhat limited but useful analogy would be what happens when you vigorously shake an oil/vinegar mixture for the dressing on your salad.
- Fat is processed very differently from amino acids and sugars because of its insolubility in aqueous solution. The fatty acids are not permitted to remain unesterified because as such they could act as detergents forming micelles and disrupt essential membrane structures.

Additional questions for Chapter 4

1. Write a series of chemical reactions connecting CO_2, H_2CO_3, HCO_3^-, CO_3^{2-} by steps involving H_2O and H^+.
2. Explain what happens to the pH of the stomach epithelial lining cells as HCl is secreted into the stomach lumen.
3. Predict how the body would deal with a synthetic molecule that does not have the expected chemical linkages holding groups together? Consider the fat substitute olestra which has fatty acids esterified to a sugar.
4. The stereochemistry of sugars is very important. Give one example where two sugars differ only in the nature of their stereochemistry, but have drastically different physiological functions.
5. What structural components of bile acids make them good emulsification agents?
6. In what ways is the structure of a chylomicron similar to and different from the structure of a cell?
7. What is the chemical nature of the group that is esterified to cholesterol in the formation of cholesterol ester?
8. Propose an intermediate that could serve to couple the endergonic esterification reaction forming cholesterol esters with the exergonic hydrolysis of ATP.

Chapter 5

....

Preliminary outline of fuel distribution and utilization by different tissues of the body

Chapter summary

This chapter describes the mechanisms and regulation of the transport and distribution of food molecules. The storage of food under conditions where it is in excess of the immediate needs of the body is described, and the differences between sugar, lipid, and protein storage are pointed out. A comparison of the energy needs of five main tissue groups—liver, skeletal muscles, brain, adipose, and blood cells—serves as a source of examples for the various relationships between these tissues. Contrasted with the storage of food molecules, the response of the system to starvation conditions is described. The relationships between the metabolic needs of the five cell types are complementary. The roles that the hormones glucagon, insulin, and epinephrine play in regulating food metabolism as well as in the condition of diabetes are presented.

Learning objectives

❑ The four specific purposes of metabolism.

❑ The storage forms of sugars, fats, and proteins.

❑ The relative stored amounts of each of the three classes of food molecules.

❑ The energy needs of the five main tissue groups discussed in this chapter.

❑ Contrast the changes that occur during starvation conditions and at times of excess food.

❑ The special energy requirements of the brain and red blood cells.

❑ The nature of the metabolic relationships between the five cell types discussed.

❑ The role of the hormones, glucagon, insulin, and epinephrine, in regulating food metabolism.

❑ The role of ketone bodies in the system's response to starvation and diabetes.

❑ The effect on the system of the lack of insulin caused by diabetes.

A walk through the chapter

The purpose of metabolism

Metabolism in general can be divided into four specific purposes: the oxidation of ....(1) for energy; the conversion of precursor molecules into new cellular material; the processing of ....(2) products; and, in some instances, the generation of ....(3).

Energy storage systems

Since the body must store food for use between meals a number of specific systems have been developed. Glucose (a monosaccharide) is stored as a large polymer called ....(4). The amount of glycogen that can be stored is ....(5). Fat is stored as droplets or globules in ....(6) tissue which is specialized for this purpose. Unlike glycogen, the amount of fat that can be stored is ....(7). The storage of energy as fat is advantageous in a number of ways, as these more highly ....(8) molecules can release more energy per unit mass; thus even glucose taken up in excess of the available glycogen storage capacity will be converted to and stored as ....(9). It is interesting that, for humans, there is no net conversion of fatty acids back to ....(10).

Answers: (1) food; (2) waste; (3) heat; (4) glycogen; (5) limited; (6) adipose; (7) unlimited; (8) reduced; (9) fat; (10) glucose.

Amino acids

Amino acids are not stored as such in animals unless one considers eggs and milk, which have very specialized storage functions. Amino acids taken up in excess of immediate needs are converted to other molecules. Complete the figure below showing the branching in the pathways of amino acid metabolism.

Complete the following

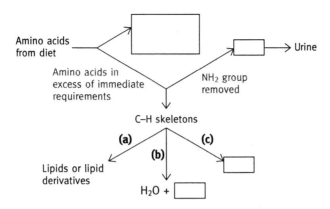

Although as a class the structures of the amino acids are quite diverse, they all contain an NH_2 group, which can be removed and used to make urea, and C—H skeletons, which can be further catabolized ($CO_2 + H_2O$) or used as the precursors for anabolic formation of members of the other classes of food substances. One important aspect of amino acid storage is that proteins that were synthesized with a specific use in mind (such as muscles) can be sacrificed if there is a more urgent need for amino acids to make another protein.

Differences between tissue types

The liver is a key player in controlling blood ....(1); however, glucose cannot be stored in quantities of more than about a ....(2) hour supply. After this period, protein is converted to glucose by a process called gluconeogenesis, in order to supply the ....(3) and red blood cells since these tissues cannot use fatty acids directly. Also, under conditions of starvation, ketone bodies are formed from ....(4) and can partially meet the energy needs of the brain. The brain has no significant fuel reserves and uses glucose as its primary food source. This requirement is of paramount importance for the organism, since without fuel brain function is significantly impaired. Skeletal ....(5) cells are major energy consumers and can utilize most types of energy sources. Adipose cells store ....(6), either directly from the free fatty acids or after being formed from glucose. The glycerol part of the triacylglycerol for storage of fatty acids comes from ....(7). Similarly to the brain, red blood cells can utilize only glucose. The five cell types are interconnected in their need to be able to mobilize the stored fuel in one cell type for use in another.

Answers: (1) glucose; (2) 24-; (3) brain; (4) fatty acids; (5) muscle; (6) fat; (7) glucose.

Complete the following

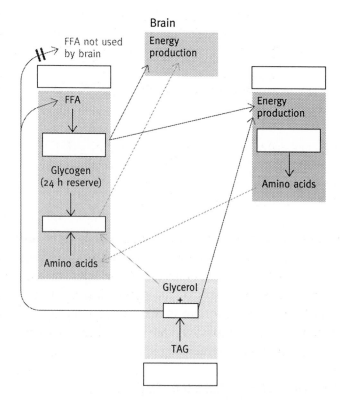

Hormones

The activity of fuel storage and use must be coordinated between the various tissue types in a manner that is appropriate for the condition in which the organism finds itself. ....(1) play a major role in conveying this information. Insulin release by the ....(2) signals that times are ....(3) and that excess fuel should be ....(4) for future use. Glucagon is released from the pancreas to signal that the level of fuel available to the cells via blood sugar is ....(5). The system's first response to glucagon is to release glucose from its storage in the ....(6), but after 24 hours when this is used up, the fat reserves in the ....(7) tissues are tapped and some muscles will begin to ....(8) and release their ....(9). The concentration of ketone bodies is ....(10) under these conditions.

In the case of diabetes, the patient lacks ....(11) and, in addition to the lack of stored glycogen, muscle cells do not use glucose as an energy source under these circumstances since the ....(12) of glucose by these cells is insulin-dependent. Extreme levels of ketone bodies can be generated due to the large amounts of ....(13) being mobilized from the adipose tissue. Epinephrine, also called ....(14), plays a role in situations where an emergency dictates the need for a very rapid availability of energy (fight or flight). ....(15) released from the nerve endings stimulates both the liver and adipose tissues to make more energy available.

Answers: (1) Hormones; (2) pancreas; (3) good; (4) stored; (5) low; (6) liver; (7) adipose; (8) degrade; (9) amino acids; (10) elevated; (11) insulin; (12) uptake; (13) fatty acid; (14) adrenaline; (15) Epinephrine.

Review of problems from the end of Chapter 5

- Saving sugars as glucose would save a lot of trouble, since it takes energy to convert glucose monomers into glycogen, however, it is not done since H_2O crosses the membrane rather freely, the cell must balance the number of individual particles dissolved per unit volume (osmolality) between the inside and outside of the cell. Many individual glucose monomers can be stored as one glycogen polymer.

     The osmotic pressure corresponds to the force generated by water molecules trying to move across the membrane. Too high a pressure, with water molecules moving *in,* will cause the cell to burst. Too low a pressure with water molecules moving *out,* will cause the cell to collapse. In order for cells to store enough glucose in the monomer form, they would have to be substantially bigger.

- TAG is a richer, more compact form of energy storage. Since TAG self-aggregates into droplets that exclude water, there is not the same problem with osmotic pressure as with other food storage forms.

- Glucose → fatty acids.   Yes, via acetyl CoA.

  Fatty acids → glucose.   No, since acetyl-CoA cannot be converted to pyruvate.

  The special case of protein storage in the form of mother's milk serves the extra needs of a developing newborn.

- The liver is of central importance in the storage and mobilization of glucose. There are three main reaction sequences involving glucose: storage and release of the monomer; synthesis of the monomer; and the conversion into fat.

- The brain uses only glucose for energy generation.

- The chief metabolic characteristic of fat cells concerns fat storage and release.

- Red blood cells use glucose and are similar to brain tissue in their relationship to energy metabolism.

- The main hormonal controls of food movement are insulin, glucagon, and epinephrine.

Additional questions for Chapter 5

1. Name the five tissue-types described in this chapter and compare them according to their respective roles in energy metabolism.
2. Why is the term 'ketone bodies' a misnomer?
3. In what sense is the metabolism of food an oxidation reaction?
4. What circumstances would lead an organism to oxidize food to produce heat?
5. What are the main fat molecules that are stored in adipose cells?
6. What happens to those amino acids that are taken up by the body in excess of the need to synthesize cellular proteins?
7. What is the difference between glucagon and glycogen?
8. What are the metabolic consequences of a fight or flight emergency situation?

Chapter 6

....

Biochemical mechanisms involved in food transport, storage, and mobilization

Chapter summary

This chapter describes the chemical reactions that are involved in the storage and use of food molecules with an emphasis on those pertaining to sugars and fats. The chemical structure of glycogen and the enzymes involved with its formation and breakdown are highlighted with regard to whether the glucose is going to be used or transported and how tissue specificity can be important. The movement of fats with a focus on TAG is followed through the five major classes of lipoprotein molecules. Coordination of the regulation of the use of sugars and fat is presented on the enzyme level and the role of cholesterol in the development of atherosclerosis is mentioned.

Learning objectives

- ❏ The direction of energy flow in the formation and breakdown of glycogen.

- ❏ The structural features of the linkages between the glucose monomers in the polymer glycogen, including the role of chain branching.

- ❏ The reactions involved in converting a glucose molecule which is part of the glycogen polymer to glucose-1-phosphate which is ready to enter glycolysis.

- ❏ The differences and similarities between the glucokinase enzyme found in liver and that form of the enzyme found in brain.

- ❏ A general understanding of the mechanism by which the system deals with sugars other than glucose.

- ❏ The abbreviations, relative compositions, and physiological functions of the five major lipoprotein particles: chylomicrons, VLDL, IDL, LDL, and HDL.

- ❏ The chemical reaction catalysed by a lipase.

- ❏ The role played by the liver in the mobilization of fats for use as energy.

- ❏ Three classes of molecules that are synthesized from cholesterol.

❏ The reaction catalysed by the enzyme LCAT.

❏ The role of the enzyme CETP.

❏ The difference between the popular medical terms, 'good' versus 'bad' cholesterol.

❏ The nature of atherosclerosis.

❏ The role of serum albumin as a transport protein.

A walk through the chapter

Glycogen synthesis

Energy must be used to store the monomeric ....(1) molecules as glycogen. Please note that some editions of the main text include a typographical error in the first sentence of the second paragraph of Chapter 6. The value of $\Delta G^{0\prime}$ for the hydrolysis of the glycosidic bond in glycogen should be negative (-16 kJ mol^{-1}). The hydrolysis is favourable energetically (with $\Delta G < 0$) so that the formation of the glycosidic bond is not favourable alone (with $\Delta G > 0$) and thus requires an input of ....(2) to occur.

The chain linkages in glycogen involve an oxygen between ....(3) and carbon 4. Carbon 1 is known as the ....(4) since in the ring open configuration this carbon has the ....(5) oxidation state; since aldehydes can be easily oxidized they are reducing agents (see review of background chemistry). The chain grows by adding groups to ....(6), making new $\alpha(1 \rightarrow 4)$ linkages with the initial reducing end always attached to the core. ....(7) is the source of energy driving the synthesis of glycogen, and hexokinase and glucokinase are two ....(8), from different tissues, that cata-lyse this reaction. Kinases are a class of enzymes that transfer a ....(9) group from ATP to the substrate. The phosphate from ATP is first transferred to carbon 6 and subsequently to carbon 1 by the enzyme ....(10). The fact that glucose-1-phosphate (G-1-P) is formed by a two-step process is part of the way in which the cell controls the synthesis of glycogen separately from its breakdown. G-1-P is then converted to the compound UDP-glucose (UDPG) by the enzyme ....(11) pyrophosphorylase using the high-energy phosphate compound UTP (which is similar to ATP). UDPG is an active form of the monomer glucose which can be added to the growing glycogen chain, a reaction catalysed by the enzyme glycogen synthase. Hydrolysis of the ....(12), abbreviated PP$_i$, produced in the synthesis of UDPG releases energy, which helps drive the overall process to completion. In addition, there is a branching enzyme, which periodically transfers a group of sugars to the oxygen of carbon 6, thus creating a branch at which additional sugars can be added.

Answers: (1) glucose; (2) energy; (3) carbon 1; (4) reducing end; (5) aldehyde; (6) carbon 4; (7) ATP; (8) enzymes; (9) phosphoryl; (10) phosphoglucomutase; (11) UDP-glucose; (12) pyrophosphate.

Complete the following

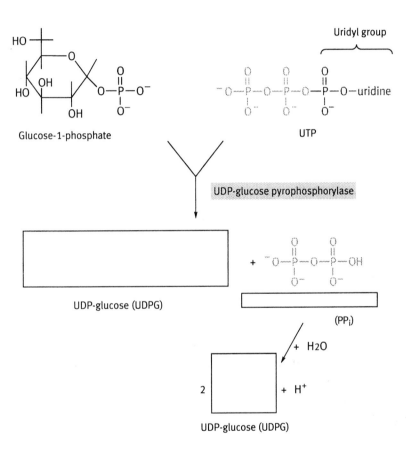

Complete the following

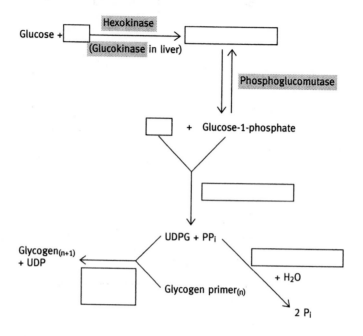

Glycogen breakdown

The mobilization of the monomer glucose from the polymer glycogen is catalysed by the enzyme glycogen phosphorylase and uses an inorganic phosphate to form glucose-6-phosphate which is converted to glucose by the actions of the enzymes phosphoglucomutase and glucose-6-phosphatase in a series of reactions that involve glucose-1-phosphate as an intermediate. There is a debranching enzyme that is also needed in this sequence. The following figure shows the relationships involved in the steps between the monomer glucose and the polymer glycogen. Note especially the opposing effects of insulin and glucagon.

Complete the following

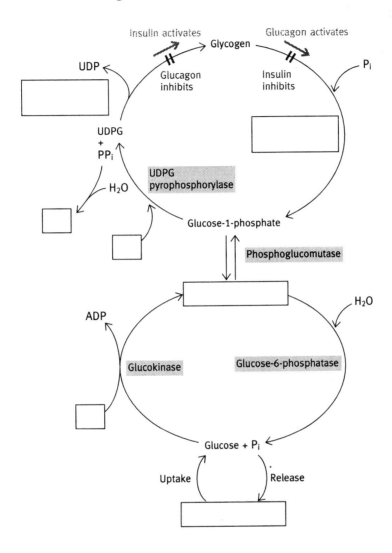

Special considerations

To keep the liver from competing with the ....(1), especially during times when the liver is in the process of converting amino acids into glucose, the liver enzyme glucokinase has a much ....(2) affinity for its substrate (glucose) than does the brain enzyme hexokinase, which performs the same function. The activity of the hexokinase is much greater than the activity of the glucokinase at the same glucose concentration. Another difference between these two enzymes that catalyse the same reaction is that hexokinase (the brain enzyme) is inhibited by

....(3), while glucokinase (liver enzyme) is not; this allows the liver enzyme to better keep up with the rate of glycogen formation. Sugars other than glucose that are digested are transformed into glucose or related compounds before being used by the organism. The sequence of events for the conversion of galactose into glucose involves phosphorylation of galactose by a ....(4), uridylation by a ....(5), with UDP-glucose as the source of –P-uridine, and epimerization of UDP-galactose via an epimerase that changes the ....(6) of the OH on carbon 4 so that UDP-galactose becomes ....(7) which can then fit nicely into the general scheme of things, since donation of the –P-uridine moiety in the transferase reaction just mentioned produces ....(8).

Answers: (1) brain; (2) lower; (3) glucose 6-phosphate; (4) kinase; (5) transferase; (6) stereochemistry; (7) UDP-glucose; (8) glucose-1-phosphate.

Fats

The movement of fats that are used for fuel involves other fatty nonpolar lipid molecules, particularly ....(1). This lipid which is essential, but for reasons other than as a fuel, is transported by the same system as the ....(2), abbreviated TAG. Chylomicrons contain, along with some proteins, phospholipids, TAG, and cholesterol esters. The fatty acids are released from the TAG by the action of a lipase called lipoprotein lipase acting outside of the cell. The free fatty acids are then rapidly taken up into the cell. The amount of ....(3) activity present outside the cell determines how actively the fatty acids are taken in. After losing much of their TAG, chylomicrons are changed into chylomicron ....(4) and taken up by the ....(5). This process is responsible for the transfer of the cholesterol esters from the ....(6) to the liver. The TAG that remained from the chylomicron remnants along with TAG synthesized in the liver from glucose and other metabolites along with cholesterol are packaged into a ....(7), which is the abbreviation for a very low-density lipoprotein particle; this is reintroduced into the circulation.

   The study of the metabolism of cholesterol is a very active one due to its importance in animal cell membranes and the correlation between high levels of cholesterol and ....(8) disease. There are also a number of important compounds that are synthesized from ....(9), for example, bile acids, steroid hormones, and cholesterol esters. Including chylomicons, there are ....(10) different lipoproteins found in circulation and each of these has its own particular set of ....(11) with various roles including involvement in the production of lipoproteins, destination-targeting, and the activation of essential enzymes.

Answers: (1) cholesterol; (2) triacylglycerols; (3) lipase; (4) remnants; (5) liver; (6) intestine; (7) VLDL; (8) cardiovascular; (9) cholesterol; (10) five; (11) apolipoproteins.

Lipoproteins

The percentage compositions of the various components of the lipoprotein particles are unique. As TAG is removed from the circulating ....(1), the percentage of cholesterol and its esters rises causing an ....(2) in the density. This is consistent with a decrease in size, as one would expect a marble to be more dense than a basketball. The VLDLs pass through an intermediate density lipoprotein particle (IDL) and finally become low-density lipoprotein particles (LDL). The latter are taken up into extrahepatic tissues by ....(3) mediated endocytosis. There is, in addition to this, a mechanism involving high-density lipoprotein particles (HDL) in which cholesterol is ....(4) transported from various tissues to the liver.

The alcohol (—OH) group on cholesterol can be esterified with a fatty acid to form a ....(5). This reaction is catalysed within the HDL particles by a step that involves the fatty acyl group of the phospholipid lecithin (also known as phosphatidylcholine) being transferred to cholesterol. This is catalysed by the enzyme lecithin cholesterol acyltransferase (LCAT). The actual reverse transport of the cholesterol occurs by a sequence of events in which the cholesterol esters in the ....(6) are transferred to other lipoprotein particles in a reaction catalysed by the enzyme cholesterol ....(7) transfer protein (CETP). Since the other lipoprotein particles are destined to arrive back at the liver, the cycle of reverse transport from various cells to the liver is complete.

The popular medical terms, 'good' and 'bad' cholesterol refer to the ....(8) and ....(9) particles, respectively. High levels of LDL are associated with the occurrence of ....(10) involving the development of ....(11) containing cholesterol which can obstruct the flow of ....(12). In contrast, high levels of HDL are associated with a decrease in the risk of this disease and may be involved in the removal of cholesterol from vascular deposits.

Free fatty acids are released from adipose cells by a lipase, which is activated by glucagon and epinephrine while inhibited by insulin. The free fatty acids released directly from the adipose tissues into the blood are carried by the highly abundant protein, ...(13).

Answers: (1) VLDL; (2) increase; (3) receptor-; (4) reverse; (5) cholesterol ester; (6) HDL; (7) ester; (8) HDL; (9) LDL; (10) atherosclerosis; (11) plaques; (12) blood; (13) serum albumin.

Complete the following

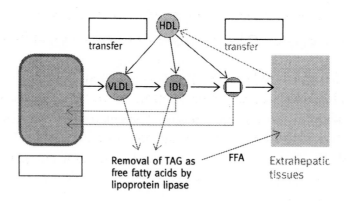

Review of problems from the end of Chapter 6

- UTP hydrolysis is coupled to the polymerization of glucose, making the formation of glycogen thermodynamically feasible.
- The term pyrophosphorylase in an enzyme's name implies that the substrate is pyrophosphate and this is unlikely.
- The liver plays a very important role in glucose mobilization.
- Both enzymes catalyse the same reaction, but the brain enzyme has a higher affinity (lower K_m) which is in line with the higher priority of this tissue.
- Galactose is an important precursor molecule in the synthesis of a number of metabolites; individuals who are not fed galactose can still function normally because of an epimerase enzyme that can interconvert UDP-glucose and UDP-galactose.
- The free fatty acids must be removed from the TAG before they can enter the cell.
- VLDL is just one of the many members of the class of lipoprotein particles.
- Reverse cholesterol flow refers to the movement from the tissues to the liver.
- The process of removing cholesterol involves chemical transformations that make it more water-soluble.
- Fatty acids, which are already esterified, can be transferred to form an ester with another alcohol without the need for a large input of energy. In this case a fatty acid ester from a phospholipid acts as the donor.
- The release of fat from adipose tissue does not occur via the TAG molecule but must first involve hydrolysis of TAG to the free fatty acid.

Additional questions for Chapter 6

1. Describe in general terms how sugars other than glucose are digested.
2. What other compounds can be synthesized from cholesterol?
3. How many fatty acids can be hydrolysed off a TAG molecule by a lipase enzyme?
4. What role does serum albumin play in food transport?
5. What chemical reaction does a mutase enzyme catalyse?
6. What is it about the names hexokinase and glucokinase that makes them so similar? (Other than the fact that they are both kinases.)
7. Name two instances in which glucagon and insulin act in complementary ways.
8. What is the structural difference between glucose and galactose?
9. What constitutes a pair of epimers?
10. Why is cholesterol not classified as an 'energy-yielding' food molecule?
11. Trace the path of a chylomicron particle.
12. Estimate the diameters in angstroms of chylomicrons, VLDL, LDL, and HDL.
13. How does the percentage of TAG compare among the various lipoprotein particles?
14. What can activate the adipose cell hormone-sensitive lipase?

Chapter 7

Energy production from foodstuffs—a preliminary overview

Chapter summary

This chapter describes the overall strategy of energy generation from the three main classes of food molecules: sugars; lipids; and proteins. With the oxidation of glucose as the main pathway, the three phases of this process—glycolysis, the citric acid cycle, and electron transport—are compared. The importance of the reduced coenzymes NADH and $FADH_2$ is examined and the role of reduction potentials, E_0', in determining overall reaction energetics is highlighted using the Nernst equation. The cellular locations of the three phases are presented along with the way in which the oxidation of fatty acids and amino acids fits into the main pathway. The fact that some organisms can convert fatty acids into glucose while others cannot is explained.

Learning objectives

- The overall energetics of glucose oxidation.

- The three phases of energy production: glycolysis; the citric acid cycle; and electron transport.

- The direction of electron transfer in an oxidation reaction.

- The difference between the oxidized and the reduced forms of the two cofactors NADH and $FADH_2$.

- The importance of hydrogen atoms in balancing the redox reactions of cofactors.

- The number and kind of energy equivalents formed in each of the three phases.

- The effects that aerobic versus anaerobic conditions have on the recycling of NADH, especially in the case of muscles.

- The cellular location of the citric acid cycle, and the unique membrane structure of the mitochondrion.

- The role that coenzyme A plays as an acetyl group carrier.

❑ The steps involved in the conversion of pyruvate into acetyl-CoA.

❑ The cellular location of the electron transport system.

❑ The roles played by O_2 and H_2O in the transport of electrons from glucose oxidation.

❑ The method of half-reactions and the calculation of coupled $\Delta E_0'$ values from half-reaction reduction potentials.

❑ The chemical nature of the thiol ester bond.

❑ Calculation of $\Delta G^{0'}$ from $\Delta E_0'$ using the Nernst equation.

❑ The manner in which the oxidation of fatty acids and amino acids can be integrated into the scheme of overall energy production.

❑ The relationship between the interconversion of glucose and fatty acids in different organisms.

A walk through the chapter

Overall strategy of energy generation

Energy is derived from glucose via oxidation in a reaction that essentially represents combustion since oxygen is the oxidizing agent and the products are ....(1) and ....(2). The reaction is extremely exothermic. (Some editions of the main text contain an error in the sign of $\Delta G^{0'}$; it is -2820 kJ mol^{-1}.) The entire process can be divided into three phases: glycolysis which occurs in the ....(3) of the cell; the tricarboxylic acid cycle (TCA), also known as the Krebs or ....(4), which occurs inside mitochondria in eukaryotes; and the electron transport system which occurs in the ....(5) mitochondrial membrane in eukaryotes.

Oxidation is the process of ....(6). As glucose is oxidized it loses its electrons to ....(7). As oxygen accepts the electrons from glucose it becomes reduced to ....(8). The transfer of electrons from glucose to oxygen occurs in a series of individual steps involving a large number of intermediate compounds known as electron carriers. Two of these are highlighted in this chapter—nicotinamide adenine dinucleotide (NAD$^+$) and flavin adenine dinucleotide (FAD). Because NAD$^+$ is a small organic molecule that participates in enzyme reactions it is known as a coenzyme. NAD$^+$ cycles between its reduced form (carrying two electrons) and its ....(9) form. Electrons are often transferred with an accompanying hydrogen atom in a species known as the hydride (H$^-$) ion. The reduced form of this coenzyme carrying the hydride is written ....(10). In chemical reactions the reduced form is accompanied by a proton, NADH + H$^+$, to balance the two hydrogens (one as H$^+$ and one as H$^-$), which are also lost as well as the two electrons from the species being oxidized. FAD is converted to ....(11)

when it is carrying electrons (as is the related compound FMN which converts to $FMNH_2$). These electron carriers play similar roles to that of $NAD^+/NADH$, but with some important distinctions that will become more important later.

Answers: (1) CO_2; (2) H_2O; (3) cytoplasm; (4) citric acid cycle; (5) inner; (6) losing electrons; (7) oxygen; (8) H_2O; (9) oxidized; (10) NADH; (11) $FADH_2$.

Glycolysis

The first stage in the production of energy from glucose, called ....(1), constitutes a series of reactions in which the six- ....(2) sugar is broken down into two molecules of the three-carbon metabolite ....(3). There is also a net production of two molecules of ....(4) and two molecules of the reduced cofactor ....(5). The NADH produced needs to be changed back to NAD^+ for the cycle to continue and one of two mechanisms occur, depending on the availability of ....(6). When plenty of oxygen is present (...............(7) conditions) the mitochondria of eukaryotes can perform this function. When there is a little or no oxygen available (.........................(8) conditions), which can often occur within highly active muscle cells, the pyruvate is reduced to lactate by the enzyme ....(9). The lactate thus formed is recycled into other metabolic processes discussed later in the text.

Answers: (1) glycolysis; (2) carbon; (3) pyruvate; (4) ATP; (5) NADH; (6) oxygen; (7) aerobic; (8) anaerobic; (9) lactate dehydrogenase.

Complete the following

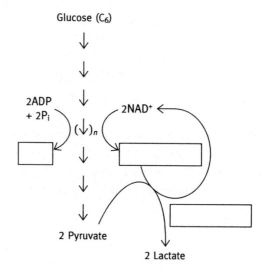

In yeast a similar anaerobic sequence produces acetaldehyde from pyruvate followed by ethanol production in a process that re-oxidizes the NADH to NAD^+. This reaction is known as fermentation. If glycolysis is performed on the polymer glycogen rather than on the monomer glucose, as is often the case, the net production of ATP is three instead of two.

Citric acid cycle

The second stage of glucose oxidation, the citric acid cycle, occurs in the mitochondria, which are double-membrane organelles with an inner and outer membrane. Pyruvate from the first step is transported across the inner membrane to the matrix region.

Pyruvate is converted into the highly versatile intermediate acetyl-coenzyme A (acetyl-CoA) before entering the TCA cycle. Coenzyme A is a cofactor which carries acetyl groups via a thiol ester linkage (see organic chemistry review). The conversion of pyruvate to acetyl-CoA involves three molecular events: the decarboxylation of pyruvate, the formation of one NADH and the coupling of this two-carbon-containing molecule from the decarboxylation of pyruvate to coenzyme A. The rest of the TCA cycle completes the oxidation of the carbons from glucose to CO_2 and in the process produces three reduced molecules of NADH, one reduced molecule of $FADH_2$, and one high-energy phosphate compound for each acetyl group fed into the cycle.

Complete the following

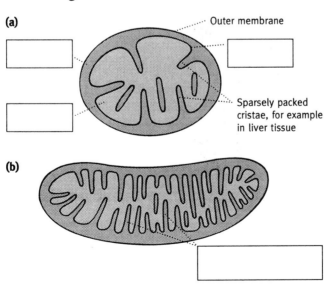

(a)

Outer membrane

Sparsely packed cristae, for example in liver tissue

(b)

Electron transport

The third stage of energy production from glucose is called the electron transport system and occurs in the inner mitochondrial membrane. The net reaction involves the transfer of electrons from the reduced (electron-carrying) cofactors (NADH and $FADH_2$) to oxygen (O_2) in a process that results in the formation of water. Reactions that involve the transfer of electrons can be separated into half-reactions showing the individual steps and separating the reactants that gain the electrons from the reactants that lose the electrons. Every redox reaction has one of each type of half-reaction.

Electrochemistry (review)

One can keep track of the energy associated with these individual and combined steps with the use of the redox potential, E_0'. E_0' values can be applied to either a half-reaction in which a single species is gaining or losing electrons, or they can be applied to the whole reaction in which two species are involved, one losing and one gaining electrons. It is common practice to compare E_0' values for half-reactions based on the reaction in question written as a reduction in which the single species involved gains electron(s) and becomes reduced. Accordingly, the smaller (more negative) the E_0'(red) value, the lower the affinity of the reagent for the electron and the less likely is the reduction reaction to occur. Thus electrons flow spontaneously (downhill in energy) from the nicotinamide cofactor with E_0'(red) $= -0.32$ V to oxygen with E_o(red) $= 0.82$ V, since O_2 has a greater affinity for the electrons than does the nicotinamide cofactor.

In calculating the free energy change associated with this transfer of electrons the Nernst equation is used along with a value for E_0' that represents the coupled half-reactions. To do this the E_0'(red) values for the half-reactions are added but with a change of sign on the one representing the oxidation half of the reaction. For the reaction in question between the nicotinamide cofactor NADH and oxygen O_2, it is the NADH that is getting oxidized so the sign of the E_0''(red) for this species must be changed before it is added to the E_0'(red) value for the O_2 half-reaction. Note that in some early editions of the main text there is an error in this section. The correct calculation for the E_0' is:

$$E_0' = 0.32 \text{ V} + 0.81 \text{ V} = 1.136 \text{ V}.$$

This positive E_0' value for the combined half-reactions gives a very negative (favourable) value for $\Delta G^{0'}$ in the Nernst equation (-219.25 kJ mol^{-1}). A very fine background description of the way in which E_0' values are measured is given in the main text. The Nernst equation is a mathematical expression that relates the change in redox potential of an electrochemical couple to the change in free energy (and thus to the equilibrium constant). Upon application of this equation to the transfer of electrons from NADH to O_2 one finds that a great deal of energy is given off. This energy is released in a series of steps that are coupled to the production of ATP from ADP and P_i.

Complete the following

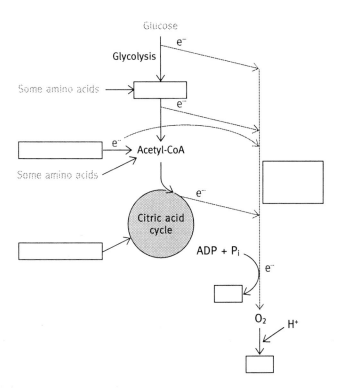

The role of fats

The oxidation of fat (mainly the fatty acid portions of TAG) and amino acids to generate ATP is nicely coupled with the three phases of the oxidation of glucose. The ....(1) from TAG are broken down two carbons at a time and attached to ....(2) to be fed into the citric acid cycle. The conversion of the long carbon chain of fatty acid into two-carbon pieces also results in the generation of NADH and $FADH_2$ to be fed into the electron transport phase of the oxidation of glucose. The oxidation of each of the 20 amino acids requires the common step of ....(3). The remaining carbon skeletons (different for each amino acid) are converted either to pyruvate or acetyl-CoA or to another intermediate in the citric acid cycle. The fact that excess glucose can be changed and stored as ....(4) is a result of the fact that acetyl-CoA can be used to synthesize fatty acids. However, animals cannot (in a net conversion) make fatty acids into ....(5) because acetyl-CoA cannot be converted to ....(6). Organisms that can do this operate by a cycle known as the ....(7).

Answers: (1) fatty acids; (2) coenzyme A; (3) deamination; (4) fat; (5) glucose; (6) pyruvate; (7) glyoxylate cycle.

Complete the following

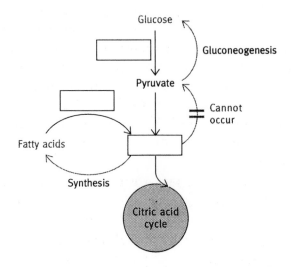

Review of problems from the end of Chapter 7

- Make a table to remember these ideas.

Phases of energy production	Location
Glycolysis	Cytoplasm
Citric acid cycle	Mitochondrial matrix
Electron transport	Inner mitochondrial membrane

- The structure of the nicotinamide coenzyme consists of a business end with a six-membered ring containing a nitrogen and a $CONH_2$ attached. In the reduced form it has two hydrogens attached to the carbon on the ring opposite the nitrogen and two double bonds in the ring. In the oxidized form it has only one hydrogen on this carbon with three double bonds in the ring which makes the nitrogen positively charged. With regard to hydrogen transfer, biochemists often refer to two hydrogen atoms (H•) each with one electron. This takes place because the reduced form always appears in the chemical reaction balanced with another H^+.

- The $FAD/FADH_2$ system plays a very similar role to that of the $NAD^+/NADH$ system.

- Aerobic means in the presence of O_2, while anaerobic means in its absence. Muscles often need to use glucose faster than oxygen can be supplied to this tissue. In doing so they must function anaerobically and the pyruvate formed is converted to lactate in order for there to be NAD^+ available to continue this process.

- Coenzyme A has an —SH group as the business end. This thiol reacts in a similar manner as a hydroxyl —OH, forming esters with carbonyls. However, the thiol esters formed between CoA and a carbonyl are of a higher energy potential than would be the case for a simple oxygen ester. This corresponds to a more negative $\Delta G^{0\prime}$ of hydrolysis. The reason for this is that the sulfur atom is bigger than oxygen and its electron orbitals do not overlap as well as do the orbitals of oxygen with those of carbon (which is also small). This means that the bond between the sulfur and carbon in a thiol ester is less stable than the corresponding O—C bond in a simple ester. This less stable molecule is thus a higher energy compound.

- The importance of the pyruvate dehydrogenase reaction stems from the fact that it connects glycolysis to the citric acid cycle by the very versatile metabolite, acetyl-CoA.

- Acetyl-CoA links glycolysis and the citric acid cycle.

- The two reduced cofactors NADH and $FADH_2$ are treated in a similar fashion to ultimately generate ATP energy.

- E_0' values are generally reported for reduction reactions. To calculate the E_0' for the coupled system we must change the sign of the half-reaction that represents the oxidation and add the E_0'. The correct equation is

$$E_0' = 0.816 \text{ V} + 0.19 \text{ V} = +1.035 \text{ V.}$$

Some earlier editions of the book have a typographical error in the answer to this question. The $\Delta G^{0\prime}$ for the oxidation of $FADH_2$ by oxygen is most definitely negative and favourable

$$\Delta G^{0\prime} = -nFE_0' = -194.06 \text{ kJ mol}^{-1}.$$

- Acetyl-CoA is the metabolite that connects a number of pathways.

- Glucose → fatty acids   Yes.
  Fatty acids → glucose   No, in animals; yes in bacteria.

Additional questions for Chapter 7

1. Explain in terms of the free energy content of the products versus the reactants why reactions with $\Delta G^{0\prime} < 0$ are spontaneous.

2. How much energy would be released if glucose were oxidized completely to CO_2 and H_2O?

3. What molecule is reduced as glucose is oxidized? Write a balanced half-reaction showing the product of this step.

4. Balance the following reaction with H^+s.

$$NAD^+ + 2e^- \rightarrow NADH + H^+;$$
$$FAD + 2e^- \rightarrow FADH_2;$$
$$\tfrac{1}{2}O_2 + 2e^- \rightarrow H_2O.$$

5. What is the difference between aerobic and anaerobic conditions?

6. What is unique about the mitochondrial membrane system?

7. Under what conditions is lactate the main product of glycolysis? Give two examples.

8. What is the chemical structure for the acetyl group attached to CoA?

9. What enzyme converts pyruvate to acetyl-CoA?

10. What are the products of the reaction that converts pyruvate to acetyl-CoA?

11. What is the role of the salt bridge in the apparatus for measuring the redox potential?

12. If a reaction in which two electrons are transferred has a net cell potential of $\Delta E_0' = 0.25$ V, what is the value of $\Delta G^{0\prime}$? Is this reaction favourable?

13. The reaction $2Fe^{3+} + 2e^- \rightarrow 2Fe^{2+}$ has a $\Delta E_0' = 0.77$ V and the reaction $\tfrac{1}{2}O_2 + 2H^+ + 2e^- \rightarrow H_2O$ has $\Delta E_0' = 0.82$ V. Will $2Fe^{2+}$ be oxidized to $2Fe^{3+}$ in the presence of O_2? (Note that the reason for the 2 in front of the Fe^{2+} and Fe^{3+} is to balance the number of electrons being transferred in each half-reaction, $n = 2$).

14. Why would dieting be easier if our bodies could effect a net conversion of acetyl-CoA into pyruvate?

15. Where do amino acids fit into the overall strategy of metabolism?

16. List the places in the cell where the three main phases of glucose oxidation take place.

Chapter 8

....

Glycolysis, the citric acid cycle, and the electron transport system: reactions involved in these pathways

Chapter summary

This chapter describes the pathways of glycolysis, the citric acid cycle, and electron transport. The details of the chemical structures and the names of the enzymes are presented. In addition, the pyruvate carboxylase reaction is mentioned as well as the coenzyme biotin. The intermediates in the electron transport chain and the generation of the proton gradient as well as the mechanism of ATP generation are presented. The numbers that relate the ATP equivalence for each reduced cofactor are explained.

Learning objectives

❑ The difference in energy requirements and yield depending on whether glycogen or free glucose is used as the starting material for glycolysis.

❑ The names and structures for each of the intermediates in the glycolysis pathway and citric acid cycle.

❑ The names of the enzymes involved in the glycolysis pathway and citric acid cycle.

❑ The terms applied to the functional group transformations that occur in glycolysis and citric acid cycles.

❑ The steps of glycolysis and the citric acid cycle that produce high-energy molecules.

❑ The reaction of pyruvate carboxylase that keeps the supply of oxaloacetate sufficient and uses biotin.

❑ The names of the major electron carriers in the generation of ATP from the reduced cofactors.

❑ The importance of coenzyme Q, also known as ubiquinone.

❑ The role played by the proton gradient across the mitochondrial membrane.

❑ The translocase mechanism for the movement of ATP and ADP.

❑ The reaction catalysed by ATP synthase.

❑ The amount of ATP that can be made from each reduced coenzyme via the electron transport mechanism.

A walk through the chapter

Glycolysis

Glycolysis, (*glyco* = sugar) + (*lysis* = splitting), involves the splitting of the six-carbon sugar glucose into two three-carbon molecules of pyruvate. Glycolysis may begin from the monomer glucose or its polymeric form glycogen. It is more energy efficient to start with glycogen, since the energy from one ATP molecule must be 'invested' to initiate the glycolysis process when starting from glucose. Each of the metabolites in the glycolysis cycle has a chemical name and many of these are often abbreviated. Each step can be described according to the chemical transformation that takes place and there is a unique enzyme for catalysing each step.

Complete the following

Metabolite	Abbrevia-tion	Structure	Reaction (enzyme name)
Glucose	Not normally abbreviated	?	Phosphorylation (hexokinase)
Glucose-6-phosphate	G-6-P		Rearrangement (phosphoglucose isomerase)

Metabolite	Abbreviation	Structure	Reaction (enzyme name)
Fructose-6-phosphate	F-6-P	CH_2OH $C{=}O$ CH_2OH CH_2OH CH_2OH $CH_2OPO_3^{2-}$	Phosphorylation (phosphofructokinase-1)
Fructose-1,6-bisphosphate	F-1,6-P	?	Breakage of C-C bond (aldolase)
Dihydroxyacetone phosphate	DHAP	?	Rearrangement (triose phosphate isomerase)
Glyceraldehyde-3-phosphate	G-3-P	?	Oxidation and phosphorylation with P_i (glyceraldehyde-3-phosphate dehydrogenase)
1,3-Bisphospho-glycerate	1,3-BPG	?	Substrate-level phosphorylation of ADP (phosphoglycerate kinase; in reverse)
3-Phosphoglycerate	3-PG	?	Rearrangement (phosphoglycerate mutase)

Metabolite	Abbreviation	Structure	Reaction (enzyme name)
2-Phosphoglycerate	2-PG	?	Dehydration to a double bond (enolase)
Phosphoenolpyruvate	PEP	?	Substrate-level phosphorylation of ADP (pyruvate kinase; in reverse)
Pyruvate	Not normally abbreviated	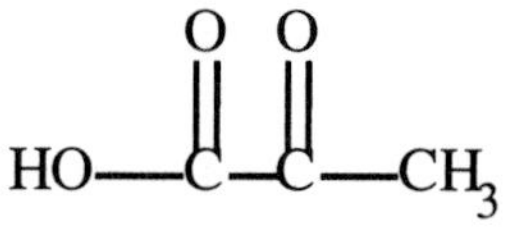 $HO-\overset{\overset{O}{\parallel}}{C}-\overset{\overset{O}{\parallel}}{C}-CH_3$	Decarboxylation and coupling to coenzyme A (pyruvate dehydrogenase) in preparation for entry into the citric acid cycle

The following notes about the individual steps will help to explain how the cycle operates. Note that the last four steps occur twice for every glucose molecule that enters the cycle. The transformation of G-6-P into fructose-6-phosphate sets the molecule up for the reverse aldol reaction (see organic chemistry notes). The fructose-6-phosphate is transformed into the 1,6-bisphosphate in order that both of the three-carbon pieces will contain a phosphate. The specific conditions inside the cell create a situation in which there is not a good correlation between $\Delta G^{0'}$ and the actual ΔG value due to the actual concentrations of the reactants and products being very different from standard conditions (1 M concentrations) and the fact that two products are being formed from one reactant molecule. The next reaction is an isomerase so that both of the three-carbon pieces can follow the same next step in which the aldehyde is oxidized to a phosphate ester. The phosphate from this molecule (1,3-bisphosphoglycerate) is transferred to ADP to form ATP by a kinase. This reaction is known as a substrate-level phosphorylation. The product 3-phosphoglycerate is then rearranged by a transfer of the phosphate from carbon 3 to carbon 2, a water molecule is removed by the enzyme enolase, and the phosphate is transferred to ADP (generating ATP) by a kinase (the enzyme was named as if it were working in reverse). From glucose there has been a net gain of two ATP molecules (from glycogen a net gain of three). The NADH formed is reoxidized either aerobically by the mitochondria or anaerobically. The aerobic process involves the transfer of the electrons from NADH into the mitochondria by either the glycerophosphate shuttle or the malate–aspartate shuttle, with the latter being more energy-efficient.

Complete the following

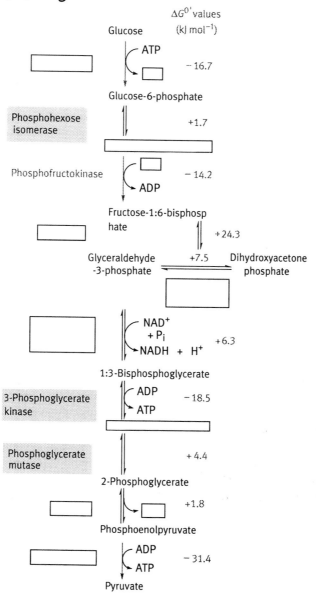

The two pyruvate molecules formed via glycolysis are transported into the mitochondria where they are first converted to acetyl-CoA in a process known as oxidative decarboxylation. The process involves a number of cofactors, including thiamin pyrophosphate and lipoic acid, and is catalysed by a multienzyme complex.

Complete the following

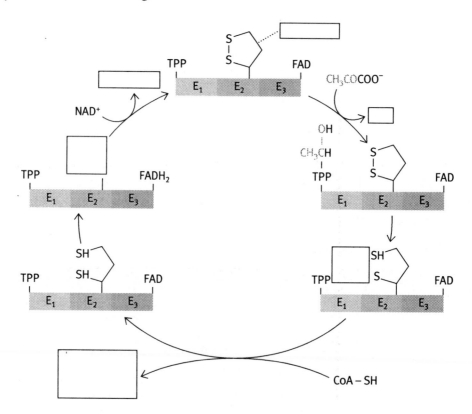

Citric acid cycle

The two-carbon acetyl group attached to its carrier coenzyme A now enters the citric acid cycle to be converted to CO_2 with the generation of three NADH, one $FADH_2$, one GTP, and the regeneration of CoASH (the un-acylated cofactor). Water is also consumed.

Complete the following

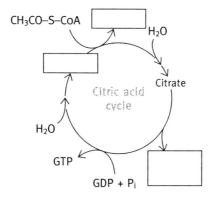

There are nine chemical structures (counting the intermediate *cis*-aconitate) and eight enzymes involved in this cycle. Oxaloacetate can be thought of as the beginning and ending point of the cycle. The following list of reactions outlines the steps of the citric acid cycle.

Complete the following

Metabolite	Structure	Reaction (enzyme name)
Acetyl-CoA	$CoA-S-\overset{\displaystyle O}{\overset{\|}{C}}-CH_3$	Joined to oxaloacetate to form citrate (citrate synthase)
Oxaloacetate	?	Joined to acetyl-CoA to form citrate (citrate synthase)
Citrate	?	Rearrangement to form *cis*-aconitate (aconitase)
Cis-aconitate	?	An intermediate in the rearrangement of citrate to form isocitrate

Metabolite	Structure	Reaction (enzyme name)
Isocitrate	?	Oxidative decarboxylation to form α-ketoglutarate (isocitrate dehydrogenase)
α-Ketoglutarate	?	Decarboxylation and coupling to coenzyme A to form succinyl-CoA (α-ketoglutarate dehydrogenase)
Succinyl-CoA	?	Hydrolysation to succinate (succinyl-CoA synthetase; operating in reverse)
Succinate	?	Oxidation to fumarate, an alkene (succinate dehydrogenase)
Fumarate	?	Stereospecific hydration (fumarase)
Malate	?	Oxidation: alcohol to ketone (malate dehydrogenase)
Oxaloacetate	?	Back to the beginning of the cycle

In summary, the two carbons of acetate are added to oxaloacetate to form citrate, which is rearranged to isocitrate and decarboxylated to α-ketoglutarate with the production of one NADH. α-Ketoglutarate is decarboxylated and the remaining four-carbon piece is transferred to CoA with the product of one NADH. The succinyl-CoA thus formed loses the CoA generating GTP from GDP and P_i, and the resulting succinate is dehydrogenated to fumarate, generating one $FADH_2$, followed by hydration to give malate and oxidation yielding one NADH and the product oxaloacetate, which can then begin another cycle.

The intermediates that make up the cycle do not exist in isolation from the rest of metabolism and are involved in a give and take equilibrium (described later). One important reaction that keeps the supply of oxaloacetate readily available is the transformation of pyruvate by the enzyme pyruvate carboxylase, an enzyme that uses ATP and HCO_3^- and requires biotin as a cofactor.

Complete the following

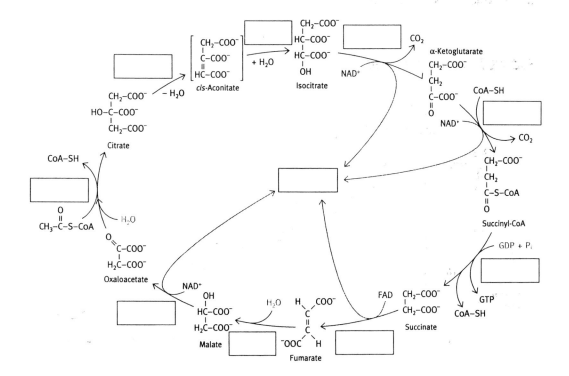

Electron transport

The final stage 3, the sequence that deals with the molecules of NADH and $FADH_2$ that have been formed, occurs in or on the inner mitochondrial membrane. There are a number of individual electron carriers involved in the overall sequence. The cytochromes contain the heme prosthetic group. The heme contains iron, which oscillates between the Fe^{2+} ferrous and Fe^{3+} ferric forms. There are also non-heme electron carriers containing iron–sulfur centres, which accept electrons from the $FADH_2$ and $FMNH_2$ and transfer them to coenzyme Q (also known as ubiquinone), which is freely soluble and mobile in the nonpolar phase of the membrane, and which operates as a one-electron carrier via the semiquinone free radical. There are also members of this group of electron carriers that are fixed in a membrane. Electrons are transported in the energetically favourable direction from molecules that easily give them up to molecules that readily accept them. The process can be described as the passage of electrons from molecules with a low redox potential to molecules with a high one. This free energy is released ($\Delta G < 0$) as the process occurs.

Complete the following

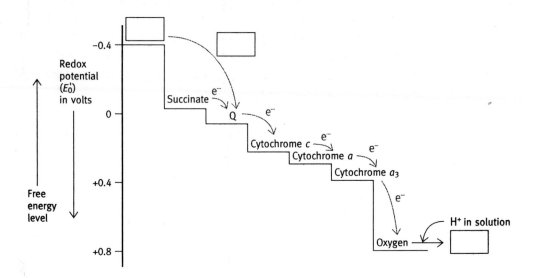

Complete the following

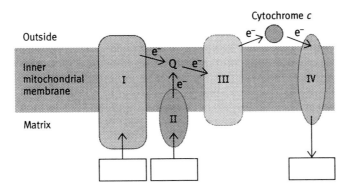

The electrons that are transferred are eventually delivered to oxygen which forms water in a reaction that also involves H^+. Coupling of the energy released as the electrons move from one carrier to the next with the formation of ATP is accomplished by generating a proton gradient pumping the protons (H^+) outside the membrane and then allowing the flow of protons back across this membrane to drive an enzyme called the ATP synthase complex, which couples ADP + P_i to form ATP in an overall process that is known as the chemiosmotic mechanism. The mechanism of how the protein gradient is produced is not completely understood but what is known about the details of the role that Q plays in complex III suggests that, as Q becomes QH_2, the protons picked up come from the matrix (inside) of the inner mitochondrial membrane and, when QH_2 is regenerated as Q, the protons are released on the outside, and thus the gradient is created.

Complete the following

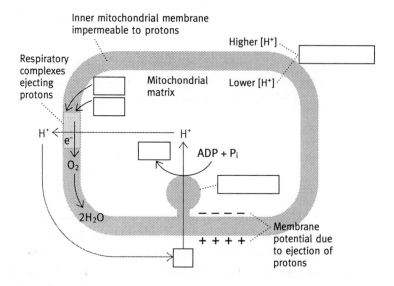

The generation of ATP involves conformational charges in the protein driven by the flow of protons back into the matrix. A good analogy is to think of the ATP synthase as a proton pump acting in reverse. Instead of using ATP to pump protons, it is letting the protons flow and making ATP. A transport mechanism, known as ATP–ADP translocase, exists to move these energy-carrying cofactors from the mitochondria where the ATP is generated to other regions of the cell where the ATP is used. Much of the energy for this is also coupled to aspects of the movement of protons across these membranes. It takes three protons to flow through the ATP synthase to generate one molecule of ATP, and one additional proton to translocate these cofactors. Thus four are needed to make one ATP. The number of protons that are pumped in response to an NADH or $FADH_2$ giving its electrons to oxygen is dependent on a couple of factors. For NADH the number can be 2.5 or 1.5 and for $FADH_2$ it is 1.5. Starting from free glucose (rather than glycogen) the net yield of ATP is either 30 or 32 depending on which shuttle is used for the cytoplasmic NADH. *E. coli* do not contain mitochondria but operate a similar pathway using the cell membrane. There are other uses of the energy stored in the generation of a proton gradient such as the brown fat cells in newborn babies which generate heat, and certain types of ion pumping and cilia rotation in some bacteria.

Review of problems from the end of Chapter 8

- Phosphohexose isomerase in effect moves the carbonyl from carbon 1 in glucose to carbon 2 forming fructose. The reaction that follows this is catalysed by aldolase, which requires this ketone carbonyl.

- Substrate-level phosphorylation means that the phosphate that is transferred comes from one of the substrate metabolites, as compared to using P_i as the phosphate source. There are two examples in the glycolysis pathway and one in the citric acid cycle.

- The name for the enzyme pyruvate kinase is based on the fact that ATP is involved in the reverse direction. The energetics of the system drives the reaction in the direction forming ATP because phosphoenolpyruvate is a high-energy intermediate, in fact higher than ATP.

- More ATP energy is generated in glycolysis starting from glycogen rather than from glucose because one ATP must be used to activate the monomer glucose while an inorganic phosphate can be added and a monomer removed from glycogen in a single step that does not require ATP. Note that some early editions of the text contain a typographical error in the answer to this question. The correct answers are two and three, respectively.

- There are two membrane shuttle pathways for NADH, with the malate–aspartate shuttle being more efficient than the glycerophosphate pathway. Part of the difference in the amount of energy generated from the reduced cofactors stems from the way in which NADH and $FADH_2$ are handled in the electron transport chain.

- The missing H_2O in the citric acid cycle comes from the reaction,

$$GDP + P_i \rightarrow GTP + H_2O,$$

   which is often abbreviated without showing the H_2O.

- Oxidation of isocitrate forms an α-keto acid with the carbonyls on adjacent carbons. The structure can then lose CO_2 and form the thiol ester with coenzyme A in a single step. In some early editions of the main text there is a typographical error that states incorrectly that the product is a β-keto acid.

- Anaplerotic reactions refer to those reactions that interconvert metabolites. In the citric acid cycle, the amount of oxaloacetate available is crucial so that incoming acetyl-CoA molecules can be brought into the cycle. If conditions are such that there is not enough oxaloacetate, then it can be formed from pyruvate in an ATP-dependent carboxylation using biotin as the cofactor.

- Biotin is a CO_2 carrier.

- The arrangement of the respiratory complexes in the electron transport chain involves four complexes, all contained in the inner mitochondrial membrane. Electrons flow from complex I to complex IV.

- Small organic molecules function to transport electrons. Ubiquinone is lipid-soluble while cytochrome c is water-soluble.
- The work that is obtained from the transport of electrons is realized in the generation of a proton gradient across the inner mitochondrial membrane.
- The difference in the values for the yield of ATP comes about because there are two possible membrane transport systems for NADH, with one being more efficient than the other. In *E. coli* there are no intracellular membranes for things to be transported across.

Additional questions for Chapter 8

1. What reaction couples glycolysis with the citric acid cycle?
2. What are two other common names used to refer to the citric acid cycle?
3. If glycolysis is a linear cycle and the citric acid cycle is a circular cycle, what term would describe β-oxidation?
4. List the enzymes of glycolysis at which energy is produced.
5. List the enzymes of the citric acid cycle that produce energy.
6. What is the nature of the pH difference between the inside and the outside of the mitochondria?
7. What is the structural difference between glucose and fructose?
8. What useful role does the enzyme phosphoglycerate mutase play in the overall design of the glycolysis pathway?
9. Using the values for $\Delta G^{0\prime}$ for each of the steps in glycolysis, which are the most exergonic steps and what do they all have in common?
10. What is one major difference between the mitochondrial glycerol-3-phosphate dehydrogenase and the cytoplasmic form?
11. Why is there not a transport system for the outer mitochondrial membrane?
12. What is the unique chemical linkage that is different between the oxidized and reduced forms of lipoic acid?
13. What is the name of the citric acid cycle intermediate with five carbons? What amino acid is similar to this compound?
14. What is the functionality of the carbon of oxaloacetate becoming attached to the acetyl group in the reaction catalysed by citryl synthase?
15. At what steps are CO_2 molecules released from the citric acid cycle? What do these steps have in common?
16. How many molar equivalents of H^+ are produced in the overall citric acid cycle?
17. Propose a reasonable structure for the intermediate in the reaction,

$$CO_2 + \text{biotin} + ATP \rightarrow \text{biotin}—CO_2 + ADP + P_i.$$

18. At which stage of the metabolic oxidation is the greatest amount of ATP produced?

19. What is the charge on the oxidized form of the iron in heme?

20. What is the characteristic structural feature that identifies the semiquinone cofactor?

21. Explain what is meant by the phrase 'electrons are transferred from a carrier of higher reducing potential to one of lower reducing potential'.

22. Explain in your own words the two parts of the chemiosmotic mechanism.

23. Describe the two other physiologically important uses of the proton gradient besides the generation of ATP.

Chapter 9

....

Energy production from fat

Chapter summary

This chapter describes the chemical reactions involved in the production of energy from fat. The relationship between this pathway and glycolysis is explained as well as the role of the acyl-CoA transport system. The ways in which the system responds to a number of varied conditions, such as the overproduction of acetyl-CoA and the metabolism of fatty acids with either an odd number of carbons or an unsaturated double bond, are also included.

Learning objectives

❑ The importance of fat as a concentrated form of energy storage.

❑ The importance of acetyl-CoA as the point of convergence between fat oxidation and glycolysis.

❑ The hydrolysis reaction of TAG by which three fatty acids are released.

❑ The reactions responsible for the activation of fatty acids to acyl thiol esters.

❑ The reason a transport system is needed for the activated fatty acids.

❑ The chemical terms describing the four main reactions involved in each functional group transformation in the cycle that produces a two-carbon acetyl thiol ester.

❑ The number and type of reduced coenzymes produced at each step of the oxidation of a fatty acid.

❑ The number of times the cycle must be repeated for a fatty acid starting with a given number of carbons.

❑ The fate of the acetyl-CoAs produced by the breakdown of a fatty acid.

❑ The way in which unsaturated fatty acids are handled differently from saturated ones.

❑ The nature of conditions where acetyl-CoA production can exceed the ability of the citric acid cycle to handle this metabolite.

❑ The consequences of excess acetyl-CoA production.

❑ Two examples of chemicals that are classified as ketone bodies.

❑ An outline of the pathway involved in the metabolism of long chain fatty acids with an odd number of carbons.

A walk through the chapter

Background

Fat is a very concentrated form of chemical energy storage which can be oxidized to produce ....(1). The pathway for the oxidation of fat converges with that of glycolysis at the common intermediate ....(2). After the free fatty acids are released from the TAG by hydrolysis, the three-carbon glycerol is phosphorylated, oxidized, and fed into glycolysis as dihydroxyacetone phosphate. Free fatty acids are obtained from albumin carriers in the ....(3) or via lipase action on the lipoprotein particles, ...(4), or VLDL.

Activation

Fatty acids are activated to the acyl-CoA form. Note that the specific term 'acetyl' refers to the two-carbon piece from acetate; similarly, propionyl is the acyl group with three carbons. Fatty acids usually have between 16 and 22 carbons and, when esterified, are referred to simply as an acyl group. The activation of a carboxylic acid generally refers to the process of making an ester or anhydride. In the case of activation with coenzyme A this involves a thiol ester. It is an endergonic step to couple the carboxylic acid of fatty acid with a thiol (sulfur analogue of an alcohol). The source of this energy is the coupling of the reaction with ....(5) hydrolysis. It is, in fact, a three-step process involving a ....(6) intermediate and PP_i formation. The enzymes that catalyse this reaction are called fatty acyl-CoA ....(7). This reaction occurs in the outer mitochondrial membrane but, since the breakdown to the two-carbon acetyl-CoA occurs in the mitochondrial matrix, a transport mechanism is needed. This process involves ....(8) and a pair of transferase enzymes.

Answers: (1) ATP; (2) acetyl-CoA; (3) blood; (4) chylomicrons; (5) ATP; (6) fatty acid-AMP; (7) synthetases; (8) carnitine.

A spiral cycle

Four reactions are involved in the cleavage of each ....(1) carbon piece from the large chain fatty acid: oxidation to a compound with a ....(2)   one carbon away from the carbonyl; hydration, which puts a ....(3) group on the carbon two positions away from the carbonyl; and a second oxidation, which makes this new alcohol into a ....(4). The fatty acid which is undergoing catabolism, now has two carbonyl groups. Then a new coenzyme A molecule comes in and forms a thiol ester with this second carbonyl kicking out the CoASH along with the two-carbon acetyl group. The original fatty acid with its CoA thiol ester is now two carbons ....(5)   and ready to undergo another round of oxidation–hydration–oxidation and thiolysis (formation of a thiol ester). For each of these cleavage steps one ....(6) and one ....(7) are produced. A molecule such as palmitate with 16 carbons needs to be clipped ....(8)   times to generate ....(9) two-carbon pieces.

Answers: (1) two-; (2) double bond; (3) hydroxyl; (4) ketone; (5) shorter; (6) NADH; (7) FADH$_2$; (8) seven; (9) eight.

Complete the following

The acetyl-CoAs go into the ....(1) and then the reduced cofactors go into the ....(2) system to generate ATP. In calculating the net production of energy one must take into account that two ATPs are consumed in the initial fatty acyl-CoA activation, since the product AMP is two high-energy phosphate bonds steps removed from ATP.

Unsaturated fatty acids are a common component of diets and the cleavage of these into molecules of acetyl-CoA is very similar to that of saturated fatty acids, with the exception of an ....(3) enzyme to put the double bond in the correct position with respect to the carbonyl or, depending upon the structure of the fatty acid, a reduction (hydrogenation) to simply replace the double bond with ....(4).

Ketone bodies

Under conditions either of ....(5) or in ....(6) when fatty acid breakdown is going full speed, to try to make up for the lack of glycolysis, the production of CoA may exceed the ability of the citric acid cycle to handle this metabolite (which is also being hampered by the shortage of oxaloacetate). This excess ....(7) is the source of the high levels of ketone bodies, ...(8) and ....(9). The mechanism for the formation of acetoacetate occurs in a roundabout way involving the coupling of three acetyl-CoAs to form HMG-CoA, a precursor in cholesterol synthesis.

Answers: (1) citric acid cycle; (2) electron transport; (3) isomerase; (4) hydrogens; (5) starvation; (6) diabetes; (7) acetyl-CoA; (8) acetoacetate; (9) β-hydroxybutyrate.

Additional points

Although rare in animals, long chain fatty acids with an odd number of carbons do occur and are handled in the same manner as their even-numbered analogues until the three-carbon propionyl-CoA is produced. Notice that a 19-carbon fatty acid would produce eight two-carbon pieces plus a three-carbon residue. This three-carbon acyl-CoA is converted to the four-carbon CoA derivative, succinyl-CoA, which is used (an intermediate in the citric acid cycle) by a series of reactions involving carboxylation and stereo- and regio-chemical changes by an epimerase and mutase, respectively. Fatty acids can also be oxidized in peroxisomes which are small membrane-bound vesicles found in many animal cells.

Review of problems from the end of Chapter 9

- Make a table.

Major carriers of fatty acids	Chemical form
Serum albumin	Free fatty acids
Chylomicrons	TAG
VLDL	TAG

- Cells of the brain and red blood cells must derive their energy needs from glucose since they do not use free fatty acids for this purpose.
- (a) Fatty acids must first be activated to the acyl-CoA form before they can be metabolized.
  - (b) The outer mitochondrial membrane is the site of fatty acid activation.
  - (c) The mitochondrial matrix is the site of fatty acid breakdown.
  - (d) The transport of fatty acids involves the molecule carnitine.
- There is a three-step sequence involving a double bond, an alcohol, and a carbonyl which is very similar in both fatty acid oxidation and the citric acid cycle.
- Palmitoleic acid (containing 16 carbons) must be activated and then split into eight acetyl-CoAs (by making seven cuts); these acetyl-CoAs must run through the citric acid cycle and the resulting reduced cofactors (and GTP) must be converted into ATP equivalents.

Step	For each turn	Example for C_{16} fatty acid
β-oxidation	1 NADH	$1 \times 7 = 7$ NADH
	1 $FADH_2$	$1 \times 7 = 7$ $FADH_2$
Citric acid cycle	3 NADH	$3 \times 8 = 24$ NADH
	1 $FADH_2$	$1 \times 8 = 8$ $FADH_2$
	1 GTP	$1 \times 8 = 8$ GTP
Electron transport		*Totals:*
	2.5 ATP/NADH (max)	$31 \times 2.5 = 77.5$ ATP
	1.5 ATP/$FADH_2$	$15 \times 1.5 = 22.5$ ATP
	1 ATP/GTP	$8 \times 1.0 = 8.0$ ATP
Cost of activating fatty acid	-2 ATP	Total $= 108 - 2 = 106$ ATP

- A monounsaturated Δ^9-fatty acid has a double bond between carbons 9 and 10. This is treated in the regular fashion of β-oxidation through three steps yielding:

$$CH_3-(CH_2)n-CH\underset{cis}{=\!=}CH-CH_2-\overset{\displaystyle O}{\overset{\displaystyle \|}{C}}-S-CoA$$

This molecule is then isomerized to give the *trans*-Δ^2-enoyl-CoA which can then continue through the process of β-oxidation. Note that in some early editions of the main text there are typographical errors in the answer to this problem.

- Acetyl-CoA is not always fed into the citric acid cycle; the system needs to be more versatile. One such alternative pathway is the formation of ketone bodies.

- Cholesterol synthesis and acetoacetate synthesis occur in different regions of the cell. It is not as yet clear to scientists why these two reactions both involve HMG-CoA.

Additional questions for Chapter 9

1. In what chemical sense is fat a more concentrated storage form of chemical energy than carbohydrates?
2. How many ATP equivalents could be made from a saturated fatty acid with 22 carbons?
3. Outline the reaction involved in the activation of a fatty acid to enter the β-oxidation pathway.
4. Why is the fatty acid transport system needed?
5. List the pathways that are connected by the intermediate acetyl-CoA.
6. What are the consequences of excess acetyl-CoA production?
7. Propose an intermediate in the reaction between ATP, CoASH, and a free fatty acid to form acyl-CoA, AMP, and PP_i.
8. What is the role of the fatty acyl-carnitine molecule?
9. What is the group that is transferred by the enzyme carnitine acyltransferase?
10. What does HMG-CoA stand for? What amino acid is this similar to?
11. What is the metal ion that is part of deoxyadenosylcobalamin?
12. What happens to the succinyl-CoA that is formed from the metabolism of the three-carbon piece resulting from the oxidation of fatty acids with an odd number of carbons?
13. How is the oxidation that occurs in peroxisomes different from that which occurs in mitochondria?

Chapter 10

A switch from catabolic to anabolic metabolism—first, the synthesis of fat and related compounds in the body

Chapter summary

This chapter describes the synthesis of fats, primarily triacylglycerol, from the two-carbon precursor, acetyl-CoA. The relationship between this pathway and the catabolic steps that break down fat and the differences between the animal and bacteria systems are highlighted. The mitochondrial membrane transport system for acetyl-CoA as well as the role of unsaturated fatty acids is mentioned. The reactions involved in the pathway leading to TAG for storage are presented, and these are contrasted with the reactions of the pathway leading to glycerophospholipids. The role of the fatty acid class known as the eicosanoids is presented, as well as their physiological function. A modest description of the synthesis of cholesterol is included with reference to three mechanisms for the control of the synthesis of this molecule.

Learning objectives

- Energy is needed to drive synthetic anabolic reactions to completion.

- The similarities and differences in the pathways of fatty acid degradation and synthesis.

- The types of reactions that involve biotin as a cofactor.

- The role of malonyl-CoA in fatty acid synthesis.

- The chemical reaction describing the interconversion of free fatty acids and TAG.

- The role of the acyl carrier protein (ACP).

- The enzymes and intermediates in the synthesis of fatty acids.

- The similarities and differences between the animal and bacterial pathways of fatty acid synthesis.

- ❑ The cellular location of fatty acid synthesis and the role of citrate in the process of transporting acetyl-CoA across the mitochondrial membrane.

- ❑ An overview of the way in which the system meets its need for unsaturated fatty acids.

- ❑ The reactions involved in TAG formation.

- ❑ The roles of ATP and CTP in the formation of glycerophospholipids and similarities and the difference between this pathway and that which forms TAG.

- ❑ The three main groups of eicosanoids.

- ❑ Three mechanisms for the control of cholesterol synthesis.

- ❑ The general concepts involved in feedback inhibition.

A walk through the chapter

Fat synthesis

A switch is made in the course of the discussion relating to the direction of ....(1) flow by turning from reactions that ....(2) energy from food to the production of long chain ....(3) by coupling acetyl-CoAs together. Excess acetyl-CoA from glycolysis must be converted to ....(4) since it cannot be converted back to ....(5) in animals. It is a general principle that the catabolic and ....(6) sides of related metabolic pathways occur by different individual steps rather than by simply using the same set of enzymes running in opposite directions. This makes both pathways thermo-dynamically favourable, and thus ....(7) and separately controllable.

To begin the process, a carbon is added to the two-carbon acetyl-CoA to form ....(8), with the expenditure of a molecule of ....(9), in a reaction catalysed by acetyl-CoA carboxylase using biotin as a cofactor. This system uses a thiol ester carrying system that is different from but related to that based on coenzyme A and that is known as the acyl ....(10) protein (ACP). The three-carbon group of the malonyl-ACP is linked with another two-carbon acetyl group on this thiol ester in a reaction involving the loss of ....(11), and thus the formation of a four-carbon molecule. Both of the carboxyl groups, one from malonyl and the other from acetyl, are present in this molecule. In a series of three steps that are the reverse of those in fatty acid degradation, one of the carbonyls (the one from the acetyl) is reduced to an ....(12) followed by the loss of ....(13) to give a double-bond ....(14). This alkene is reduced to give the saturated (CH_2) at that carbon. In the next round, another malonyl thiol ester is brought in and, with an accompanying decarboxylation, the four-carbon chain becomes six carbons, and the reduction–dehydration–reduction to give the saturated carbon is repeated.

Answers: (1) energy; (2) release; (3) fatty acids; (4) fat; (5) carbohydrate; (6) anabolic; (7) irreversible; (8) malonyl-CoA; (9) ATP; (10) carrier; (11) CO_2; (12) alcohol; (13) H_2O; (14) alkene.

Complete the following

In animal tissues as opposed to the bacterial (cf. *E. coli*) system, all of the catalytic activities of the steps following malonyl-CoA formation reside on one protein. The growing fatty acid chain is never released from the enzyme complex until the fatty acid synthesis is completed. The thiol ester attachment acts as an anchor. The successive reductions to remove the ketone carbonyl which is formed involve the cofactor ....(1) which is very similar to NADH, but is involved as a reagent in anabolic steps while NADH is a product of ....(2) ones. Fatty acid synthesis (anabolic) occurs in the ....(3) as compared to fatty acid oxidation (catabolic) which occurs in the ....(4). Acetyl-CoA made in glycolysis is transported across the mitochondrial membrane via an ATP-dependent mechanism in which it is converted to ....(5), transported across the membrane, and converted back to ....(6) and ....(7). The oxaloacetate is reduced to ....(8) by NADH and the malate is oxidatively decarboxylated to ....(9) producing

NADPH. The pyruvate can be made back into oxaloacetate by a ....(10) enzyme. The citrate transport mechanism accomplishes two necessary tasks by getting the acetyl group ....(11) of the mitochondria and generating ....(12) for fat synthesis. In a sense the two electron reducing equivalents of NADH are transferred indirectly to the NADPH. Only certain of the unsaturated fatty acids that are needed by the body can be synthesized; others containing multiple double bonds (cf. linoleic acid, linolenic acid) must be obtained through the ....(13). The primary storage form of fatty acids is as the glycerol ....(14) triacylglycerol (TAG). The acyl-CoA ester of the fatty acid is transferred to the free alcohols of ....(15) 3-phosphate obtained by the reduction of ....(16) phosphate, an intermediate in the glycolysis cycle.

Answers: (1) NADPH; (2) catabolic; (3) cytosol; (4) mitochondria; (5) citrate; (6) acetyl-CoA; (7) oxaloacetate; (8) malate; (9) pyruvate; (10) carboxylase; (11) out; (12) NADPH; (13) diet; (14) ester; (15) glycerol-; (16) dihydroxyacetate.

Complete the following

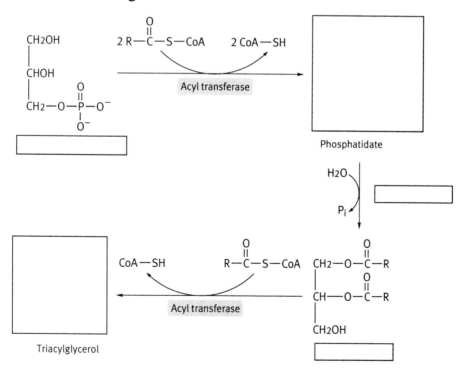

Phospholipids

A related set of compounds, the glycerophospholipids, are *not* prepared by the ester-ification of an alcohol to the phosphate of the glycerol after two fatty acids have already been attacked. The alcohol of the polar group being attached to make the phospholipid is activated in a two-step process involving ATP and CTP and is then attached to the last free alcohol of diacylglycerol. This reaction sequence is used to make phosphatidylethanolamine and phosphatidylcholine, but in the synthesis of phosphatidylinositol and cardiolipin, the phosphatidic acid (diacylglycerol + one phosphate) is activated to the CDP analogue and then the unactivated alcohol is added. In addition, there is interconversion between the different classes of phospholipids. Membrane lipid synthesis occurs on the smooth endoplasmic reticulum (ER) membrane.

Complete the following

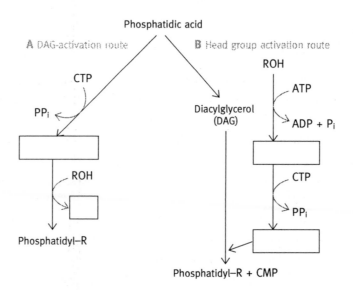

Prostaglandins and related compounds

Eicosanoids, which contain 20 carbon atoms, have a wide range of physiological functions including the ability to act as local ....(1). There are three main groups: the ....(2); ....(3); and ....(4). The main subclasses of prostaglandins are PGA, PGE, and PGF, with subscript numbers to indicate the number of double bonds in the side chain attached to the cyclopentane ring. The role of aspirin can be (at least partially) explained by its ability to inhibit a ....(5) enzyme which is the first step in prostaglandin ....(6). Leukotrienes also arise from ....(7) acid by another mechanism.

One very famous lipid (or fat) that the body can make is cholesterol which is the result of a long detailed metabolic process starting from only the simple two-carbon molecule acetyl-CoA. Of particular interest are the first few steps leading up to the formation of ....(8) acid since it is here that the control of the synthesis of cholesterol occurs. Cholesterol synthesis is controlled by three mechanisms all involving the same enzyme, ...(9). The cholesterol molecule itself can: inhibit the synthesis of this enzyme protein at the gene (DNA) level; cause the enzyme to be ....(10); and inactivate the protein by a mechanism that involves a ....(11) step. Since cholesterol is the ultimate product of the sequence of reactions involving the HMG-CoA reductase enzyme, this type of regulation is called a ....(12) mechanism. The therapeutic drugs, Lovastatin and ....(13), act to inhibit this key enzyme.

Answers: (1) hormones; (2) prostaglandins; (3) thromboxanes; (4) leukotrienes; (5) cyclooxygenase; (6) synthesis; (7) arachidonic; (8) mevalonic; (9) HMG-CoA reductase; (10) destroyed; (11) phosphorylation; (12) feedback; (13) Simvastatin.

Review of problems from the end of Chapter 10

- Malonyl-CoA is an activated form of the two-carbon piece that is added to the growing fatty acid chain. This is an example of coupling exergonic steps with endergonic ones to make the net transformation favourable.
- The reductive steps of fatty acid synthesis change the carbonyl to a methylene ($-CH_2-$).
- Multienzyme complexes are more efficient for multiple step pathways than are individual monomeric enzymes.
- One example of metabolic compartmentalization is the case in which NAD^+ is used in catabolic steps and $NADP^+$ is used in anabolic steps.
- Not all tissues specialize in fatty acid synthesis.
- Acetyl-CoA must be transported from inside the mitochondria to the cytoplasm; this transport step involves citrate.
- There are ways to convert the reducing equivalents carried by NADH to NADPH reducing equivalents. In addition, the pentose phosphate pathway supplies some of this required material.
- The general theme of an alcohol plus activated carboxylic acid to give an ester is very common in biochemistry.
- The use of CDP-conjugates as activated forms is common.
- The eicosanoids are a class of molecules with a number of very specific and rather varied functions. The effects seen after taking aspirin correlate nicely with the role this compound plays in perturbing the metabolism of eicosanoids.

- Drugs that have structures analogous to the transition state structure of a specific reaction act as inhibitors by competing with the natural substrate for the active site of the enzyme.

Additional questions for Chapter 10

1. What is the rationale behind the statement, 'the catabolic and anabolic sides of related metabolic pathways occur by different individual steps'?
2. Draw the abbreviated three-carbon compound structure of malonyl-CoA.
3. Which steps of the citric acid cycle are similar to those in fatty acid synthesis?
4. When the three-carbon malonyl-CoA is linked with the growing fatty acid chain, how many carbons are added and what happens to the rest?
5. Name two polyunsaturated fatty acids that must be obtained from the diet.
6. Name the three main groups of eicosanoids.
7. Name three mechanisms involved in the control of cholesterol synthesis.
8. Name two therapeutic drugs used to treat high levels of blood cholesterol and describe how they function.
9. What reaction does the enzyme HMG-CoA reductase catalyse and why is this important?
10. What are the names of the classes of molecules that make up coenzyme A?
11. Where is the extra phosphate that is not found in NAD^+ attached in $NADP^+$?

Chapter 11

Synthesis of glucose in the body (gluconeogenesis)

Chapter summary

This chapter describes the anabolic pathway of glucose synthesis from pyruvate. The two reactions needed to convert pyruvate into phosphoenolpyruvate are highlighted. The system's response to the starvation condition is presented and includes alternative sources for pyruvate. The importance of the Cori cycle, and the method of recycling glycerol from TAG hydrolysis, are also highlighted. The reversibility of the pathway between glucose and pyruvate is contrasted with the fact that acetyl-CoA cannot be converted to glucose in a net reaction in animals. The glyoxylate cycle that accomplishes this in plants is outlined.

Learning objectives

- ❑ The importance of glucose and the consequences of starvation.
- ❑ Pyruvate can be converted back into glucose, but, in animals, acetyl-CoA cannot.
- ❑ The thermodynamic reasons why only certain glycolysis steps are reversible.
- ❑ The enzymes involved in gluconeogenesis that bypass glycolysis steps.
- ❑ The mechanism of the Cori cycle for using lactate from anaerobic muscle activity for making glucose.

- ❑ The mechanism for using glycerol from TAG hydrolysis for making glucose.
- ❑ The types of organisms that can make sugar out of fat.
- ❑ The logic behind the idea that, through the glyoxylate cycle, acetyl-CoA can be used to make glucose in a *net* sense.
- ❑ The enzymes and intermediates of the glyoxylate cycle.

A walk through the chapter

Glucose synthesis

Glucose is the only usable food source for the ....(1)   and, since the
.....................(2)   only stores a limited (~24-hour supply) of glucose as
............................(3), glucose must be synthesized if conditions of starvation
continue past this time interval. Pyruvate (the three-carbon product of glycolysis)
can be converted back to glucose but, once pyruvate is converted to the two-carbon
acetyl-CoA for use in the citric acid cycle, glucose cannot be re-made, in a net sense.
This means that fatty acids that are reversibly connected with acetyl-CoA by either a
catabolic or ....(4) couple are *not* connected to glucose in animals.
Plants are a different story.

   Many of the reactions that convert pyruvate to glucose are similar to the ones that
operate in glycolysis, except of course the ....(5) that are ....(6)
due to thermodynamic energy barriers. The synthesis of glucose bypasses these
three steps. Pyruvate is converted to phosphoenolpyruvate (PEP) by the alternative
mechanism involving two steps and consuming one ....(7) and one ....(8).
While this is occurring, the back conversion of PEP into pyruvate is inhibited. The
other two thermodynamically irreversible reactions in glycolysis are phos-
phorylations by ....(9);   these can be bypassed by simple hydrolysis
reactions with enzymes known as phosphatases.

Answers: (1) brain; (2) liver; (3) glycogen; (4) anabolic; (5) three; (6) irreversible; (7) ATP; (8)
GTP; (9) kinases.

Complete the following

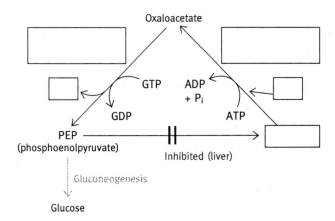

The Cori cycle

The main source of pyruvate for the synthesis of glucose after the reserves have been used is ....(1) protein. Alanine is of key importance since it is transported in the blood from the muscle to the ....(2), as well as being one of the amino acids that can be converted into ....(3) through the citric acid cycle and then into pyruvate. As a second source, there is lactate produced from the anaerobic glycolysis that occurs after strenuous muscle activity when the glycolysis rates exceed the capacity of ....(4) to reoxidize NADH. This cycle of lactate from the muscle travelling to the liver to become pyruvate for glucose synthesis is known as the ....(5) cycle; it is a way to recycle the lactate that is formed as a result of stress caused by overexertion of the muscle.

Answers: (1) muscle; (2) liver; (3) oxaloacetate; (4) mitochondria; (5) Cori.

Complete the following

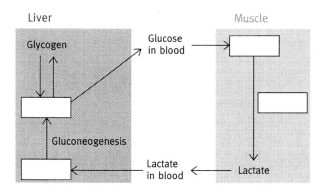

Glycerol formed from TAG hydrolysis in ....(1) tissue can be used for glucose synthesis in the liver. This involves phosphorylation by a kinase and oxidation of the alcohol to give ....(2) phosphate which can enter the cycle just described as glycolysis in reverse.

Glyoxylate cycle

The limitation of not being able to make sugar out of fats does not hold for all organisms (cf. *E. coli* and plant seeds). In addition to the normal citric acid cycle, these organisms possess the enzymes required for the ....(3) cycle which introduces a shortcut in the citric acid cycle avoiding the ....(4) steps. This is done so that those recovered carbons can effect a net gain in the metabolic intermediate ....(5) needed to effect the synthesis of glucose. Using this pathway, the acetyl-CoAs from fatty acid oxidation can be made into glucose. Glyoxylate is the two-carbon metabolite with both an aldehyde carbon and a carboxylic acid.

Answers: (1) adipose; (2) dihydroxyacetone; (3) glyoxalate; (4) decarboxylation; (5) oxaloacetate.

Complete the following

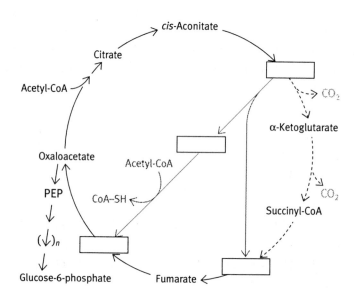

Review of problems from the end of Chapter 11

- It is always instructive to question the utility of the design of metabolic systems. There are almost always very important reasons why the system behaves the way it does. In this case, the energy metabolism of the brain and red blood cells is dependent almost exclusively on the availability of glucose.

- Phosphoenolpyruvate is a much higher energy metabolite than pyruvate. To run the reaction in the reverse direction of glycolysis, using ATP to put the phosphate back on, the enol form of pyruvate would be required and the following equilibrium lies far to the right.

$$\underset{\text{enol}}{CH_2{=}\underset{\underset{\displaystyle OH}{|}}{C}{-}CO_2^-} \quad \rightleftharpoons \quad \underset{\text{ketone}}{CH_3{-}\underset{\overset{\displaystyle O}{\|}}{C}{-}CO_2^-}$$

This step is circumvented in gluconeogenesis by a two-step process, each step using a high-energy phosphate.

- Some early editions of the book contain a typographical error in the answer to question no. 3. The two enzymes are fructose-1:6-bisphosphatase and glucose-6-phosphatase. As a class, the phosphatase enzymes catalyse the hydrolytic cleavage of a specific phosphate group.
- One effect of the phosphorylation of glucose is to keep it from diffusing back out of a cell. Muscles are never called upon to release the glucose they already have and so do not contain this phosphatase.
- In short, the Cori cycle involves lactate from the muscle being converted to glucose in the liver in order to be transported back to the muscle for use.
- Only the liver can make glucose out of glycerol. This is where the enzyme glycerol kinase is found.
- The glyoxylate cycle is operational in plants and bacteria.

Additional questions for Chapter 11

1. What are the three enzyme steps of glycolysis that are bypassed in gluconeogenesis?
2. What is the main source of pyruvate for glucose synthesis after the reserves of glucose have been used up?
3. How does the lactate produced from anaerobic muscle activity fit into an animal's need for glucose?
4. Why is it important to inhibit the step catalysed by pyruvate kinase in the liver during times of starvation?
5. Alanine from muscle breakdown is transported to the liver for glucose synthesis. What is the structural similarity between alanine and pyruvate?
6. Where does the split occur between the traditional citric acid cycle and the glyoxylate pathway?

Chapter 12

....

Strategies for metabolic control and their application to carbohydrate and fat metabolism

Chapter summary

This chapter describes the way in which the two main energy-metabolizing systems based on sugars and fats are controlled and coordinated at the enzyme level when there is excess food available, when food is not available, and in emergencies when energy is needed as fast as possible. These processes must also be coordinated among the different tissue types involved in the storage and use of these food molecules. The chapter includes discussions on the Michaelis–Menten model of enzymes with allosteric control and the role of protein phosphorylation in response to hormone activation involving a second messenger.

Learning objectives

- ❑ The overall direction of metabolic flow under conditions of feast versus famine.

- ❑ The problem posed by a 'futile' or substrate cycle and how systems are designed to avoid this.

- ❑ The thermodynamic differences among reactions that are singled out as control points.

- ❑ The two basic ways of reversibly modulating an enzyme's rate of activity.

- ❑ The factors influencing the amount of an enzyme present in the system and the concept of protein half-life.

- ❑ The importance of the Michaelis–Menten model of enzyme kinetics, and the effects of [S], K_m, and v_{max} on the rate at which an enzyme catalyses a reaction.

- ❑ The concept of feedback control.

- ❑ The importance of allosteric control of enzyme reactions and a graphical interpretation of its effect on enzyme kinetics.

❏ The two models put forth to describe the mechanism of allosteric control.

❏ The importance of phosphorylation and dephosphorylation in controlling enzyme kinetics and the amino acid side chains that can be involved.

❏ The roles of kinases and phosphatases in covalent modification of proteins.

❏ The difference between intrinsic and extrinsic control of enzyme activities and the general role of hormones and neurotransmitters.

❏ The important control points of glycogen breakdown and formation.

❏ The important control points of pyruvate dehydrogenase as the enzyme involved in the metabolic link between fatty acid breakdown, glycolysis, and the citric acid cycle.

❏ The idea of energy charge as reflected in the concentrations of ATP/ADP/AMP and NADH/NAD$^+$.

❏ The important control points of fatty acid breakdown and synthesis and the necessity of coordinating these two pathways.

❏ The effects of the hormones, glucagon, insulin, and the epinephrines, on muscle, liver, and adipose tissue and how these effects are coordinated.

❏ The importance of the second messenger cAMP and the role of cell membrane receptors in regulating the enzyme adenylate cyclase.

❏ The importance of the AMP-dependent protein kinases (PKA).

❏ The importance of amplifying the signal by using a series of kinases.

❏ The mechanisms for turning back off the activation of a system by a signal.

❏ The importance of the novel intermediate, fructose-2:6-bisphosphate.

A walk through the chapter

Controlling enzyme activity

The metabolism of carbohydrates and fats constitutes the largest flux of materials in most systems. The direction of this flux is changed very frequently since feast and famine is the way for many creatures. Balance is the goal so that the various pathways work in unison rather than in opposition. This involves mechanisms of communication.

To put it simply, if the reactions in one direction are running, the reactions in the opposite direction need to be ....(1). These various systems must respond to the need for ....(2) and the availability of ....(3), and the potential problem of ....(4) cycles or substrate cycles (analogous to an electrical short circuit) must be avoided. Reciprocal control involves having the ....(5) signal turn the one pathway ....(6) and the other ....(7) on. Recall that it is in the presence of thermodynamically ....(8) steps (large ΔGs) that control can be exerted on each of the pathways individually. A freely reversible reaction can be catalysed by the same enzyme in ....(9) and cannot be reciprocally controlled in the two directions.

Clearly, reversible control is more desirable than irreversible control for systems that need to change the direction of the response. The two basic ways of reversibly modulating the rate of an enzyme's activity are to change the amount of the ....(10) or to change the magnitude of ....(11) that the enzyme affords to the reaction. The amount of any enzyme present is a balance between how quickly the protein is ....(12) and activated and how quickly the protein is ....(13). Half-life is a measure of how long (on average) it takes for ....(14) of the protein molecule to be degraded. ....(15) are relatively short-lived with half-lives ranging from one hour to several days. This type of control is relatively ....(16) and slowly responsive on both the step-up and step-down response.

Answers: (1) switched off; (2) energy; (3) food; (4) futile; (5) same; (6) off; (7) pathway; (8) irreversible; (9) both directions; (10) enzyme; (11) acceleration; (12) synthesized; (13) degraded; (14) one-half; (15) Proteins; (16) long-term.

The Michaelis–Menten model

The activity of enzymes that are already present can be very rapidly fine-tuned. Our understanding of this mode of regulation is based on the mechanism of enzyme action put forth by ....(1) in which the enzyme (E) and the substrate (S) combine reversibly to form a complex (ES) from which the product (P) emerges and the enzyme recycles. The concentration of substrate [S] affects the rate of reaction by shifting the equilibrium concentration of the ....(2) and thus the dependence of the rate on [S] exhibits a ....(3) saturation curve. This means that, above a certain concentration of substrate [S], the enzyme is operating at its ....(4) speed and can go no faster. Under such conditions the enzyme is said to be ....(5) with substrate and operating at ....(6). Individual enzymes have different values for the substrate concentration needed to force the enzyme to operate at its v_{max} and the v_{max} value for each enzyme is generally different. In general, the ....(7) the enzyme binds with the substrate, the less substrate is needed to cause the enzyme to operate at ....(8). The number used as a measure of the tightness of the enzyme–substrate complex is called the ....(9) and is defined as the concentration of substrate [S] at which the enzyme operates at ....(10). Thus a ....(11) K_m reflects a tight ES complex. Cellular concentrations of substrates are typically in the range of K_m.

Answers: (1) Michaelis and Menten; (2) ES complex; (3) hyperbolic; (4) maximum; (5) saturated; (6) v_{max}; (7) tighter; (8) v_{max}; (9) K_m; (10) $\frac{1}{2}v_{max}$; (11) small.

Allosteric control

Usually one certain enzyme in a pathway is regulated, often the ....(1) enzyme in a pathway. Feedback control occurs when high concentrations of the metabolite at the end of the pathway are able to slow down the first reaction in the step (also known as product inhibition). The two main mechanisms whereby enzyme activities are modified without changing the amount of enzyme available are allosteric control and covalent modification. Allosteric control involves ....(2) molecules, which bind to the enzyme and either speed it up or slow it down. The most common type involve changes in the tightness with which the enzyme can bind the substrate; this is known as a ....(3) effect. Enzymes of this type often contain multiple subunits and typically more than one catalytic subunit. These subunits interact to change the response of the enzyme to the ....(4) concentration. Graphically, this becomes a ....(5) 'S'-shaped curve. Activators move the curve to the ....(6) (lower ES), and decrease the ......(7) giving the enzyme a ....(8) ES complex. Inhibitors (deactivators) move the curve to the right (higher ES) and thus ....(9) the K_m, giving the enzyme a weaker ....(10) complex.

Two models have been proposed to explain this effect. One is the ....(11) in which the individual subunits can exist in either a high-affinity state or a low-affinity state, with them *all* being in the one or the other. The two states are referred to as relaxed or tense, respectively. It is the binding of the substrate molecule that causes the switch from a ....(12) to a ....(13) affinity state. While all of the sites on each enzyme molecule are either in the high- or low-affinity site together, the various enzyme molecules are in equilibrium between the two states with the ....(14) affinity state being the more stable of the two. Thus, as the concentration of S ....(15), a great number of enzymes are shifted to the ....(16) affinity state. The binding of separate allosteric modulators influences this equilibrium between low-affinity and high-affinity enzyme conformations, by preferentially interacting with one of the two states. There is said to be a cooperative binding of the substrates as the enzymes switch from the tense to the relaxed state.

In the sequential model, all enzyme molecules or subunits start out in the low-affinity state and there is no equilibrium but, as ....(17) binds with a substrate molecule, it changes conformation from T (tense) to R (relaxed), and this facilitates a similar change in an ....(18) subunit. It is this facilitating of the change from T to R that causes the sigmoidal curve. This process continues through all of the subunits. Both models give satisfactory explanations of the mechanism, which may, in fact, vary between or be some combination of the two.

Allosteric control can be exerted by molecules that are distinct from the substrate or product of the reaction, and this allows a very versatile method of connecting metabolic pathways. The metabolites of one pathway can be the regulators of another and enzymes can have more than one effector.

Answers: (1) first; (2) ligand; (3) K_m; (4) substrate; (5) sigmoidal; (6) left; (7) K_m; (8) tighter; (9) increase; (10) ES; (11) concerted model; (12) low-; (13) high-; (14) low-; (15) increases; (16) high-; (17) each subunit; (18) adjacent.

Control by phosphorylation

Each pathway needs to know what the other is doing to avoid 'chemical chaos' resulting in metabolic pile-ups and shortages. Phosphorylation is a form of covalent modification of an enzyme in order to change its activity. Kinases are enzymes that transfer a phosphate from ATP to another molecule; this often causes the protein being phosphorylated to change activity via conformational changes in the protein structure. The process is chemically reversible, and the phosphate is removed in a reaction catalysed by an enzyme called a phosphatase. The hydroxyl groups of the amino acids, serine, threonine, and sometimes tyrosine, are used to form the high-energy phosphate ester bond.

Complete the following

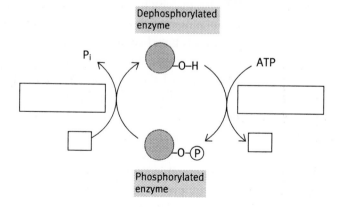

Intrinsic and extrinsic levels of control

Controlling the kinases and phosphorylases that attach and remove the phosphate groups (respectively) is often accomplished from outside the cell in response to the action of ....(1) and ....(2). Intrinsic controls tend to act via ....(3) mechanisms and operate automatically with a level of self-regulation. The coordination of the efforts of different ....(4) is the goal of the extrinsic mechanisms. The interconversion of glycogen ....(5) and phosphorylated glucose ....(6) is catalysed by a synthase and phosphorylase, respectively. The ....(7) that forms glycogen is controlled extrinsically in coordination with the availability of excess food. The ....(8) that forms glucose-1-phosphate exists in two forms: the less active phosphylase *b* and the more active phosphorylase ....(9). The inactive phosphorylase ......(10) can, however, be partially activated by the allosteric action of AMP which often attains a high concentration when energy reserves are low. This partial activation by ....(11) is undone (antagonized) by ....(12) and ....(13), which reflects the fact that, when the concentrations of these species are high, the system does not need to make more energy from glycolysis. The occurrence of both the phosphorylase *b* and *a* forms allows ....(14) control mechanisms to override the ....(15) ones.

Answers: (1) hormones; (2) neurotransmitters; (3) allosteric; (4) cells; (5) polymers; (6) monomers; (7) synthase; (8) phosphorylase; (9) *a*; (10) *b*; (11) AMP; (12) ATP; (13) glucose-6-phosphate; (14) extrinsic; (15) intrinsic.

With AMP indicating the cell's need for more energy, phosphofructokinase is also ....(1) and fructose-1:6-bisphosphatase is ....(2). Fructose-6-phosphate activates phosphofructokinase which increases the level of fructose-1:6-bisphosphate and activates ....(3) ....(4), an example of feed-forward control. The activating effects of AMP and PFK are balanced by the inhibitory effect of ....(5). Citric acid, a name that is used interchangeably with citrate in most instances, also accumulates when ATP levels are high and this also inhibits ....(6) so that glycolysis stops feeding material into the ....(7). Similarly, acetyl-CoA (the material that enters the citric acid cycle) inhibits ....(8), the enzyme that leads to the formation of acetyl-CoA. Pyruvate ....(9) is activated by acetyl-CoA in order to make more acetyl-CoA from pyruvate when there is a lot of acetyl-CoA because ....(10) is the citric acid cycle intermediate that combines with acetyl-CoA to get the cycle started.

Answers: (1) activated; (2) inhibited; (3) pyruvate; (4) kinase; (5) ATP; (6) PFK; (7) citric acid cycle; (8) pyruvate kinase; (9) carboxylase; (10) oxaloacetate.

Complete the following

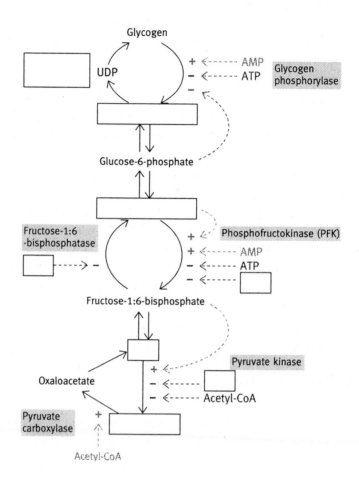

The pyruvate dehydrogenase link

Pyruvate dehydrogenase, which transforms ....(1) into ....(2), occupies a strategic control position by coupling glycolysis with the ....(3) and with fat synthesis. The products, acetyl-CoA and ....(4), inhibit the ....(5), while the substrates, ...(6) and NAD⁺, ...(7) the activity. This dehydrogenase is also subject to control by ....(8) levels through a phosphorylation step catalysed by a kinase enzyme, which is part of the whole dehydrogenase enzyme complex. The citric acid cycle and electron transport system of ATP phosphorylation are very much dependent on the availability of NAD⁺ and ADP as ....(9). Generally, a high-energy charge, measured as a high ....(10), ...(11) concentration, shuts down the energy-producing pathways, while a ....(12) energy charge, reflected in high concentrations of ADP and NAD$^+$, ...(13)   the system to charge the metabolic battery.

Answers: (1) pyruvate; (2) acetyl-CoA; (3) citric acid cycle; (4) NADPH; (5) enzyme; (6) CoASH; (7) stimulate; (8) ATP; (9) substrates; (10) ATP; (11) NADH; (12) low-; (13) stimulates.

Storage of fat

Fatty acid breakdown and synthesis are coordinated intrinsically in such a manner that they mutually suppress each other. The fatty acyl-CoA intermediates inhibit the ....(1), which serves as the first committed step in fat synthesis. The first product of irretrievably committed fat synthesis is ....(2), which inhibits the transfer of fatty acyl groups to ....(3).   Acetyl-CoA carboxylase is activated by citrate, and citrate is only transported out of the ....(4) when the level is high enough to indicate the need to store fats. The major control on the acetyl-CoA carboxylase is inactivation by ....(5) which is stimulated by ....(6). The dephosphorylation reaction that undoes this ....(7) is under hormonal control. The physiological needs of the whole body are communicated to individual cells through hormones and neurotransmitters.

Answers: (1) carboxylase; (2) malonyl-CoA; (3) carnitine; (4) mitochondria; (5) phosphorylation; (6) AMP; (7) inactivation.

Complete the following

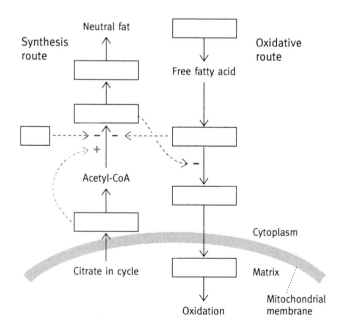

Role of hormones

The primary hormones used in the regulation of carbohydrate and fat metabolism are
...................(1), ...(2), and the catecholamines, ...(3)
and ....(4). Glucagon signals that blood sugar is ....(5),
insulin signals that blood sugar is ....(6), and both epinephrine and norepinephrine signal the body's immediate need for ....(7). Insulin is secreted from
specialized cells in the ....(8) and has only a short half-life in the blood.
The signal is thus transient and responds in a very sensitive way to blood sugar
levels above normal levels (90 mg dl^{-1}).

Hormones are chemical signals released into the blood and, while all cells are exposed to them, ...(9) respond. Hormones can be divided into two
types: ....(10) soluble and ....(11) soluble. The three discussed here are
water-soluble and interact with a ....(12) on the outside of
the cell membrane. Only ....(13) have the appropriate receptor.
The signal carried by the hormones is coupled to a ....(14),
an intracellular regulatory molecule. In the case of glucagon and the epinephrines,
the second messenger is cyclic AMP (cAMP) which is made as needed from
...........(15) by the action of the enzyme ....(16). cAMP
activates ....(17) by an allosteric mechanism, and these kinases phosphorylate the enzymes under hormonal control. These kinases as a group are called
cAMP-dependent protein kinases (PKA). In the absence of cAMP, adenylate cyclase
is a ....(18) with two catalytic and two regulatory ....(19).
The cAMP binds to the ....(20) subunits and the catalytic subunits
become fully operational. A ....(21) hydrolyses the cAMP
signal, and the PKA kinases become once again inactive. The proteins that had been
phosphorylated lose their phosphate by the action of a ....(22) and
the system returns to the resting state. The number of receptor proteins on the outside of the cell is controlled, and may decrease if the cell is overstimulated. Also,
events on the inside of the cell, such as the receptor becoming phosphorylated, can
alter the activity of the receptor. Activity here is defined as the ability of the receptor
to transfer the signal into the cell.

Answers: (1) glucagon; (2) insulin; (3) epinephrine; (4) norepinephrine; (5) low; (6) high;
(7) energy; (8) pancreas; (9) target cells; (10) water-; (11) lipid-; (12) receptor protein; (13) target
cells; (14) second messenger; (15) ATP; (16) adenylate cyclase; (17) kinases; (18) tetramer;
(19) subunits; (20) regulatory; (21) phosphodiesterase; (22) phosphatase.

Coordination of storage and use

In general, glucagon and insulin have opposite and complementary roles, although
the mechanism of insulin action is not yet completely understood. The cell takes up
glucose via a protein-mediated membrane transport mechanism, which is

..............(1) with respect to ....(2) requirements in most cells. In muscle and adipose cells this uptake is affected by levels of ....(3), while it is not affected in the ....(4). The fact that the brain enzyme ....(5) has a lower K_m (tighter binding) than the liver enzyme ....(6) for glucose has been discussed before and relates to the more important role played by glucose in the brain compared to that in the liver. ....(7) increases the activity of glucokinase by inducing the cell to make more of this enzyme protein, in response to ....(8) blood sugar levels. In the case of fat, the free fatty acids cross the membrane ....(9) due to their hydrophobicity. Thus, the amount entering the ....(10) is controlled at the same control point as that for the amount entering the ....(11).

Answers: (1) passive; (2) energy; (3) insulin; (4) liver and brain; (5) hexokinase; (6) glucokinase; (7) Insulin; (8) high; (9) unassisted; (10) cell; (11) blood.

The breakdown of glycogen

Glycogen breakdown is initiated by the enzyme glycogen phosphorylase which exists in two forms—the '*a*' which is ....(1) without the presence of ....(2) and the '*b*' which is active only in the presence of AMP. Glycogen phosphorylase *b* is converted to the *a* form by a ....(3) that phosphorylates it. In muscle, epinephrine in the blood causes this event using ....(4) as a second messenger. Epinephrine is the hormone associated with the fight or flight reaction; it has been termed the metabolic alarm button, calling for the generation of large amounts of ATP. In the liver, epinephrine causes the breakdown of glycogen to be coordinated with the release of ....(5) into the ....(6), thus maximizing the fuel supply to the muscles. Glucagon has the same effect as epinephrine on liver cells. The path of the message from the second messenger cAMP to the breakdown of glycogen is not direct, but rather involves the activation of a kinase, which activates another kinase, which activates the enzyme that breaks down glycogen. The kinase between the PKA turned on by cAMP and the glycogen phosphorylase *b* which breaks down the glycogen is called ....(7). This indirect route allows the amplification of the signal by a ....(8) process. One hormone molecule can cause the generation of several cAMP molecules and each of these can activate a second enzymatic process; the degree of enhancement at each step is multiplicative.

The mechanism for switching off the breakdown of glycogen generally involves the ....(9) of high-energy phosphate bonds by a phosphatase. In liver this enzyme is stimulated by free glucose and can be inhibited by a special inhibitor protein when that protein is phosphorylated by a ....(10) dependent mechanism. The divalent cation ....(11) also plays a role in muscle by activating the kinase that activates the enzyme that breaks down ....(12). The lowering of Ca^{2+}

concentration after the cessation of the neural signal for muscle contraction reverses this activation, a process that involves the protein ....(13).

Answers: (1) active; (2) AMP; (3) kinase; (4) cAMP; (5) glucose; (6) blood; (7) phosphorylase *b* kinase; (8) cascade; (9) hydrolysis; (10) cAMP-; (11) Ca^{2+}; (12) glycogen; (13) calmodulin.

Complete the following

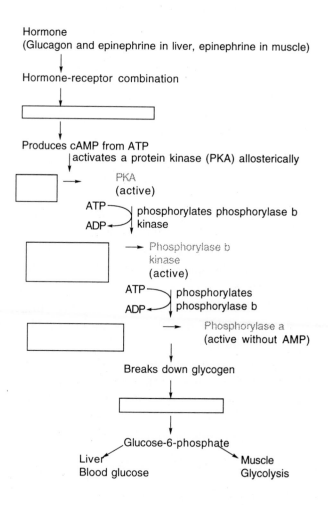

The synthesis of glycogen

Glycogen synthase is phosphorylated by the same cAMP-dependent protein kinase (PKA) as is glycogen phosphorylase *b*, but, in this case, the act of becoming phosphorylated inactivates the enzyme. Thus a dephosphorylation step by a phosphatase would activate the synthase and the protein that inhibits this phosphatase is inactivated by phosphorylation by a kinase—one that is activated directly by insulin. It is the same phosphatase inhibitor that is involved in both systems: the inhibitor is activated by a cAMP-dependent protein kinase (PKA) and inactivated by an insulin-dependent kinase.

Complete the following

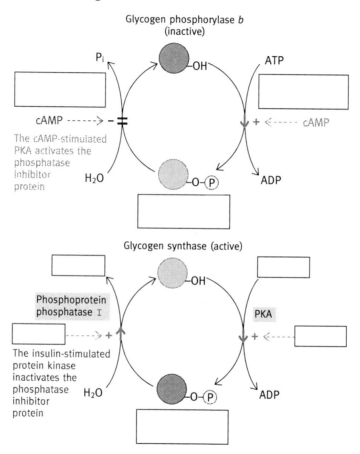

Extracellular control of glycolysis and gluconeogenesis

The hormone glucagon signals the liver to ....(1) blood glucose by either glycogen breakdown or ....(2). In liver cAMP turns on glycogen breakdown and gluconeogenesis (not glycolysis). In muscles epinephrine stimulation of cAMP production turns ....(3) glycogen breakdown but does not also ....(4), because the muscle is using the glucose-6-phosphate produced to make ATP energy available. Thus cAMP blocks glycolysis in the ....(5), but activates it in ....(6).

One important control point for glycolysis is the enzyme phosphofructokinase. cAMP inhibits this enzyme in the liver. But, in muscle, an increased concentration of cAMP caused by epinephrine results in increased glycolysis. In the liver, the effect of cAMP on phosphofructokinase (PFK) involves a metabolite not specifically involved in glycolysis, fructose-2:6-bisphosphate (note 2:6 and not 1:6). In the liver, cAMP reduces the level of fructose-2:6-bisphosphate and thus ....(7) glycolysis and activates ....(8). This is accomplished by a second phosphofructokinase (PFK$_2$), which is inhibited by cAMP by activating a kinase that phosphorylates the PFK$_2$ protein. This is an example of a double-hit effect because the ....(9)   PFK$_2$ which is inhibited from synthesizing the 2:6 compound actually ....(10) the 2:6 compound by acting as a ....(11). Note that the two reactions are not strictly the reverse of each other, as ATP is the substrate of the phosphorylation, but *not* a product of the phosphatase hydrolysis.

Answers: (1) produce; (2) gluconeogenesis; (3) on; (4) inhibit glycolysis; (5) liver; (6) muscle; (7) inactivates; (8) gluconeogenesis; (9) phosphorylated; (10) destroys; (11) phosphatase.

Complete the following

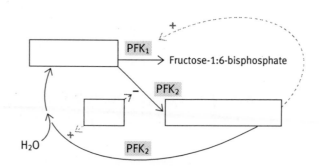

Some tissue-specific differences

In muscle the story is different since the PFK$_1$ of glycolysis must not be ....(1) when cAMP levels are increased. This process also involves the 2:6 compound and PFK$_2$ enzyme, but the details are not completely understood. Another enzyme in the glycolysis sequence that responds to cAMP is the liver (not muscle) pyruvate kinase which is ....(2) and ....(3) so that, in response to low glucose levels, glycolysis is switched ....(4). The muscle enzyme need not be regulated in this manner since it does not ....(5) glucose. The inhibition of this enzyme also allows the avoidance of the futile cycle involving oxaloacetate, PEP, and pyruvate. Thus, when pyruvate is changed into PEP (through oxaloacetate) as in the synthesis of glucose from pyruvate, the whole process isn't undone by ....(6) from glycolysis catalysing the transformation of PEP into ....(7).

In liver, the phosphofructokinase of glycolysis is ....(8) and the fructose-1:6-bisphosphatase enzyme needed for glucose synthesis from pyruvate is ....(9) so that the liver will supply the blood with needed sugar.

The control of fat metabolism is primarily at the level of the fat cells (adipose) which either store fat as triglycerides or release the free fatty acids into the blood. Insulin triggers the ....(10); glucagon triggers the ....(11). Epinephrine is also involved, as is cAMP. The lipase reaction that hydrolyses the first ester bonds of the triacylglycerol (TAG) is sensitive to phosphorylation and becomes ....(12) in response to elevated cAMP, resulting from glucagon- or epinephrine-binding to the cell receptors. Insulin works in opposition (antagonizes) these effects. The first enzyme in fat synthesis is the ....(13); this is activated by insulin and inactivated by a protein kinase activated by AMP. Glucagon or epinephrine cause the ....(14) that activates the acetyl-CoA carboxylase to be activated by cAMP. If ATP levels are low, AMP levels are high, and cause inactivation of the carboxylase and, if blood glucose is low, glucagon keeps it inactive. Insulin also ....(15) fat synthesis.

Answers: (1) inhibited; (2) phosphorylated; (3) inactivated; (4) off; (5) synthesize; (6) pyruvate kinase; (7) pyruvate; (8) inhibited; (9) activated; (10) storage; (11) release; (12) active; (13) acetyl-CoA carboxylase; (14) phosphatase; (15) promotes.

One can make a table to help keep track of the effectors that have been presented in this chapter.

Mechanism	First-order effects	Second-order effects	Overall effects
The effector cAMP			
Activates PKA	Phosphorylates proteins	Cascade amplification	Tissue-dependent
PKA activates phosphorylase kinase via phosphorylation	Phosphorylase kinase phosphorylates inactive glycogen phosphorylase *b*	Glycogen phosphorylase *b* becomes the active *a* form	Breaks down glycogen even in the absence of AMP
Inhibits phosphoprotein phosphatase by the PKA activation of a phosphoprotein inhibitor	Phosphoprotein phosphatase is inhibited from inactivating glycogen phosphorylase *a*	Glycogen phosphorylase stays active	Breaks down glycogen even in the absence of AMP
PKA phosphorylates glycogen	Phosphorylation of glycogen synthase makes this enzyme inactive		The synthesis of glycogen is inhibited

Mechanism	First-order effects	Second-order effects	Overall effects
The effector cAMP (cont.)			
Activates the phosphatase activity of PFK_2	PFK_2 phosphatase activity removes a phosphate from fructose-2:6-bisphosphate	Lower amount of fructose-2:6-bisphosphate means that phosphofructokinase is not stimulated	Glycolysis is shut down
		Lower amount of fructose-2:6-bisphosphate means that the enzyme fructose-1,6-bisphosphatase is not inhibited	Glucose synthesis can proceed
Inactivates the kinase activity of PFK_2	PFK_2 kinase is inhibited from phosphorylating fructose-6-phosphate into fructose-2:6-bisphosphate	Lower amount of fructose-2:6-bisphosphate (see above)	Glycolysis is shut down and glucose synthesis can proceed (see above)
The effector insulin			
Stimulates protein kinases	Phosphorylates proteins	Cascade amplification	Tissue-dependent
Phosphorylation of the phosphatase inhibitor keeps it from inhibiting the phosphatase	The phosphatase stays active	Glycogen synthase stays unphosphorylated and thus active	The synthesis of glycogen for glucose storage proceeds

Review of problems from the end of Chapter 12

- The importance of reversible modulation is that it can be fine-tuned to the needs of a given situation rather quickly.
- Graphically, the standard relationship between substrate concentration and reaction rate is a hyperbolic curve, while the allosterically modulated system shows a sigmoidal curve.
- The shift in the curve with the steepest portion in the region of the [S] magnifies the effects of changing concentrations.
- Allosteric modulation usually does not change the v_{max}.
- Homotropic refers to binding by the same ligand, and the cooperative effect is seen in the increase in affinity.

- The binding site on the enzyme for the ligand controlling the activity shows a different specificity than that of the binding site involved in the actual catalytic reaction.
- The distinction between intrinsic and extrinsic is based on whether the signal is completely internal to the cell or whether it is from the exterior and thus influences other cells in the whole system.
- Make a table to keep this information handy.

| Pathway | Enzyme | Modulators | |
		(+)	(−)
Glycogen breakdown	Glycogen phosphorylase	AMP	ATP, G-6-P
Glycolysis	Phosphofructokinase	Fructose-6-phosphate, AMP	ATP, citrate
	Pyruvate kinase	Fructose-1:6-bisphosphate	ATP, acetyl-CoA
Gluconeogenesis	Fructose-1:6-bisphosphate		AMP
	Pyruvate carboxylase	Acetyl-CoA	

- Pyruvate dehydrogenase is the enzyme that links pyruvate from glycolysis to acetyl-CoA which can enter the citric acid cycle. The enzyme is abbreviated PDH. There are allosteric effectors as well as phosphorylation/dephosphorylation steps, which are catalysed by a kinase and phosphatase, respectively.
- Make a table to help remember these ideas.

| Pathway | Enzyme | Modulators | |
		(+)	(-)
Synthesis of fat	Acetyl-CoA carboxylase	Citrate in plasma	AMP, fatty acid CoA
Breakdown of fat	Carnitine acyltransferase I		Malonyl-CoA

- The releases of these two complementary acting hormones, insulin and glucagon, are both triggered by the levels of blood glucose.
- Cyclic AMP is one of the best studied second messengers.
- Glucose transport is of a passive facilitated diffusion type, but the process is regulated in response to insulin binding to its receptor in the membrane.
- cAMP activates a protein kinase which activates phosphorylase *b* kinase which activates glycogen phosphorylase *b* into glycogen phosphosphorylase *a* which breaks down glycogen into glucose-1-phosphate which is converted to glucose-6-phosphate and enters glycolysis.

- Glycogenolysis is a synonym for glycogen breakdown and this is the process that is stimulated by the hormones, epinephrine and glucagon. In the liver, glucose monomers are exported to the blood; none of the monomer is used in glycolysis and, in fact, the synthesis of glucose from pyruvate is stimulated. In muscle, the glucose monomer is made available for glycolysis, the energy-yielding pathway that fuels muscle activity.

- The use of cAMP as a second messenger works even when different cell types are involved since each cell can have a unique set of receptors. The types of cells are specific for different functions when they are stimulated.

- Phosphofructokinase 1 (PFK_1) is the glycolytic enzyme that phosphorylates fructose-6-phosphate to produce fructose-1:6-bisphosphate. This enzyme is stimulated by fructose-2:6-bisphosphate and the reverse reaction of the bisphosphatase is inhibited by fructose-2:6-bisphosphate. Phosphofructokinase 2 (PFK_2) is an enzyme that can catalyse either the breakdown or formation of this modulator fructose-2:6-bisphosphate, depending on whether the protein is phosphorylated or not, respectively. The enzyme that phosphorylates PFK_2 is a cAMP-dependent protein kinase. In a bigger sense cAMP shuts down glycolysis in the liver so that the glucose that is formed from glycogen breakdown can be exported into the blood. The mechanism of this effect of cAMP involves lowering the levels of fructose-2:6-bisphosphate.

- The inactivation of pyruvate kinase in muscle would not be appropriate under fight or flight conditions because maximum muscle energy is required. It would be appropriate to do this in the liver because of the importance of generating glucose to be exported into the blood.

- The first step in mobilizing fatty acids from adipose storage tissues is to clip them off from a TAG molecule using a lipase. This enzyme is responsive to the second messenger cAMP that is formed in response to the binding of glucagon on a receptor.

Additional questions for Chapter 12

1. Why, in terms of controlling an enzyme's activity, is it better to have the enzyme operating at below v_{max}?
2. In what way could adding a phosphate to an enzyme cause changes in its structure?
3. What does it mean when we say that the activation of phosphorylase *b* by AMP is antagonized by ATP?
4. What are the three tissues that play especially important roles in carbohydrate and fat metabolism?
5. What is phosphorolysis and how is it different from phosphorylation?

6. In what sense is the regulation of glycogen phosphorylase *b* by AMP a routine metabolic control, while the conversion of glycogen phosphorylase *b* to the *a* form is a control mechanism with a higher priority?

7. Describe the action of the phosphatase inhibitor protein.

8. Complete the following table.

Effector	Mechanism	Overall effects
In muscle tissue		
Epinephrine and glucagon via cAMP	?	?
	?	
	?	?
		?
In liver tissue		
Epinephrine and glucagon via cAMP	?	
	?	?
	?	
	?	?
Glucagon only		?
Insulin	?	?
In adipose tissue		
Epinephrine and glucagon via cAMP	?	?
	?	?
Insulin	?	?

Chapter 13

Why should there be an alternative
pathway of glucose oxidation—the
pentose phosphate pathway?

Chapter summary

**This chapter describes the reactions involved in the pentose phos-
phate pathway. The cell requirements met by this process are important in understanding why this alternative pathway for glucose oxidation exists. The oxidative and non-oxidative portions of this pathway
are described and the variation in the latter part is discussed in relationship to various cell requirements. The fates of the unused secondary products of this pathway are described.**

Learning objectives

- ❑ The three specific needs of a cell
  that can be met by the pentose
  phosphate pathway.

- ❑ The products of the oxidative
  portion of this pathway.

- ❑ The two alternative non-oxidative
  routes of this pathway.

- ❑ The two products of the non-
  oxidative portion of the pathway
  and their uses in the cell.

- ❑ The role of the transaldolase and
  transketolase enzymes.

- ❑ The mechanism by which the cell
  compensates for the unequal use
  of the two products.

- ❑ To understand how sugars with
  various numbers of carbons (C_3,
  C_4, C_5, C_6, C_7) can be involved in
  converting ribose-5-phosphate to
  glucose-6-phosphate.

A walk through the chapter

In order to meet the specific metabolic needs of supplying ....(1)
for nucleotide and nucleic acid synthesis, supplying NADPH for ....(2) synthesis, and providing a route for five carbon sugars known as ....(3) to be

metabolized, a completely different pathway for glucose oxidation, the pentose phosphate pathway, exists. The pathway has two parts. The oxidation portion comprises four steps that convert ....(4) into ribose-5-phosphate; this part generates ....(5) molecules of NADPH. The second, non-oxidative part is changeable according to the ....(6) of the cell. Cells often need ribose-5-phosphate and little or no NADPH (cf. cells that are dividing) or the cell may require a great deal of NADPH and little or no ribose-5-phosphate (cf. cells actively synthesizing fatty acids from acetyl-CoA). Two enzymes, ...(7)   and ....(8), function as a pair to compensate for the unequal use of the two products. Unused ribose-5-phosphate is converted into ....(9) in a series of reactions involving C_3, C_4, C_5, C_6, and C_7 sugars.

Answers: (1) ribose-5-phosphate; (2) fat; (3) pentoses; (4) glucose-6-phosphate; (5) two; (6) needs; (7) transaldolase; (8) transketolase; (9) glucose-6-phosphate.

Complete the following

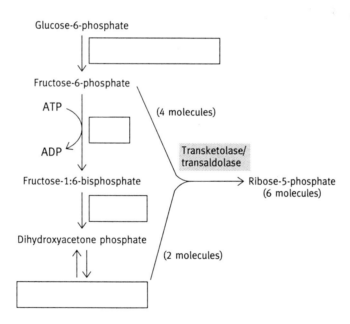

Review of problems from the end of Chapter 13

- Nucleotides contain the sugar ribose, NADPH plays a special role in anabolic steps, and normal diets contain five-carbon sugars.
- When glucose is oxidized in the pentose phosphate pathway one of its carbons goes all the way to CO_2.

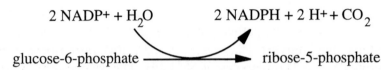

- Transaldolase and transketolase interconvert sugars in accordance with the metabolic needs of the cell.
- The net conversion for the nonoxidative reaction is

  6 Ribose-5-phosphate $\rightarrow$ 5 glucose-6-phosphate + P_i.

- If the cycle runs six times and the pentoses are converted to glucoses, the net reaction is

  Glucose-6-phosphate $\rightarrow$ 6CO$_2$ + P_i.

- NADPH has other uses in the cell besides its role in fat synthesis.

Additional questions for Chapter 13

1. How many carbons are in the piece of sugar that is transferred by the transketolase and transaldolase?
2. Are the following reactions catalysed by a transketolase or a transaldolase:
   (a) $2C_5 \rightarrow C_3 + C_7$;
   (b) $C_7 + C_3 \rightarrow C_4 + C_6$;
   (c) $C_5 + C_4 \rightarrow C_3 + C_6$?
3. What is the major structural difference between 6-phosphogluconolactone and 6-phosphogluconate?
4. What is the chemical relationship between ribose-5-phosphate and ribulose-5-phosphate? (*Hint.* The enzyme that interconverts these two is an isomerase.)
5. In what sense is the pentose phosphate pathway flexible?
6. What is the chemical relationship between ribulose-5-phosphate and xyulose-5-phosphate? (*Hint.* The enzyme is an epimerase.)
7. Using the notation that represents sugars by the number of carbons they contain, write a balanced equation that shows the conversion of fructose-6-phosphate plus glyceraldehyde-3-phosphate to give ribose-5-phosphate.

Chapter 14

··

Raising electrons of water back up the energy scale—photosynthesis

Chapter summary

This chapter describes the process of photosynthesis. The process of using energy from light to make sugar out of CO_2 and H_2O can be divided into the light and dark reactions. The role of the chloroplasts and the transport of the electrons through the various photosystems to generate a proton gradient and produce ATP are outlined. The importance of the Calvin cycle, the contrast between reaction with CO_2 and with O_2, and the unique reactions of the C_4 pathway are presented.

Learning objectives

- ❑ The role of oxidation and reduction reactions in the conversion of $CO_2 + H_2O$ to sugar and O_2.

- ❑ The two parts of photosynthesis.

- ❑ The role of the chloroplast and the constituent parts of this organelle.

- ❑ The similarities and differences between the electron transport process in mitochondria and that in chloroplasts.

- ❑ The similarities and differences between the chlorophyll and heme molecules.

- ❑ The natures of the two types of chlorophylls.

- ❑ The chemiosmotic mechanisms that generate ATP.

- ❑ The pathway used to convert CO_2 to glucose and how it differs from gluconeogenesis found in animals.

- ❑ The reaction catalysed by Rubisco involving CO_2.

- ❑ The nature of the reaction of Rubisco with O_2.

- ❑ The special ability of the C_4 synthesis pathway in some plants.

A walk through the chapter

Energy from the Sun

Photosynthesis turns the energy of the ....(1) into stored chemical energy by using high-energy electrons to reduce CO_2 into ....(2). The energy stored in these food molecules makes much of life possible because this stored energy can be released and can drive to completion biologically important chemical reactionse. When CO_2 and H_2O combine, it is the H_2O that gets oxidized to ....(3). Photosynthesis can be divided into two parts: the 'light reactions', which ....(4) H_2O and generate ATP and ....(5);   and the 'dark reactions', which use the NADPH and ATP to reduce ....(6) to carbohydrate. Light energy is directly involved only in the transfer of electrons from water to $NADP^+$ and in the generation of ATP.

Answers: (1) Sun; (2) carbohydrates; (3) O_2; (4) oxidize; (5) NADPH; (6) CO_2.

Complete the following

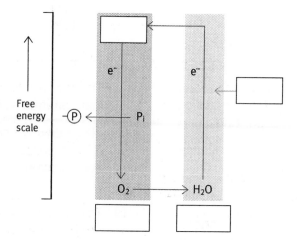

The chloroplast

The chloroplast is the site of photosynthesis. Chloroplasts are similar in many ways to mitochondria. The thylakoids of the chloroplasts of green plants contain the light-harvesting chlorophyll. The light-harvesting chlorophyll and the electron transport pathways are in the thylakoid membranes, while the conversion of CO_2 and H_2O into carbohydrate and oxygen occurs in the chloroplast stroma.

Complete the following

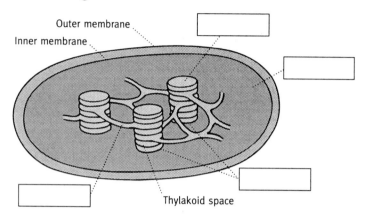

There are many similarities between the mechanism of photosynthesis and that of the electron transport chain in ....(1). There are three complexes involved. The reaction that generates NADPH is endergonic, but driven by the energy supplied by sunlight. Two ....(2) are needed per NADPH produced, and one molecule of ATP is also made. Chlorophyll is similar to the heme that carries oxygen in the ....(3), but contains ....(4) instead of iron along with other minor structural differences. The light-absorbing property of chlorophyll gives green plants their colour. There are two types of chlorophyll that work together (*a* and *b*). The direct effect of light on the chlorophyll is to excite an ....(5) to a higher energy orbital. Then, through a process known as resonance energy transfer, this energy of many excited electrons is funnelled to special reaction centres—P680 in PSII and P700 in PSI.

The sequence begins in PSII and involves pheophytin, ...(6), cytochromes, and plastocyanin. The electrons come from ....(7), pass from PSII to PSI, and are eventually carried to ....(8) in the aqueous stroma by ferredoxin. (Some early editions of the main text contain a typographical error in Fig. 14.6. The photosystem on the water-splitting centre is photosystem II and thus the labels for PSI and PSII are switched.) Four electrons are extracted from two molecules of water, and O_2 and H^+ are released into the thylakoid ....(9) by a reaction catalysed by a complex of proteins known as the water-splitting centre. ATP is generated by a chemiosmotic mechanism from ADP which is coupled to the generation of a transmembrane ....(10) potential in a manner similar to that found in mitochondria. Protons are pumped from the outside in, a direction that is ....(11) to that in the mitochondria.

Answers: (1) mitochondria; (2) photons; (3) blood; (4) magnesium; (5) electron; (6) plastoquinone; (7) water; (8) $NADP^+$; (9) lumen; (10) proton; (11) opposite.

Complete the following

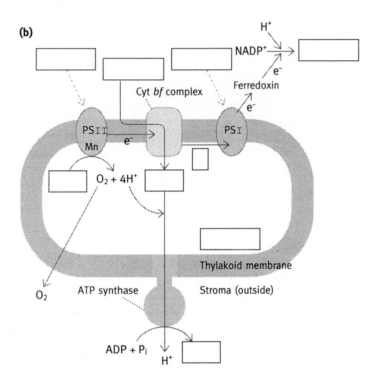

The Calvin cycle

The enzyme known as Rubisco catalyses the conversion of ribulose-1:5-bisphos-phate and CO_2 to two molecules of ....(1) by a process referred to as C_3 photosynthesis. The stoichiometry of the net conversion involves ....(2) five-carbon sugar molecules and six one-carbon CO_2 molecules which produce ......(3) three-carbon phosphorylated glycerol molecules. Two of these three-carbon glycerol phosphates are used to make a six-carbon ....(4) for storage, and the others go to make five-carbon sugars needed to continue the process. The process is complex involving sugars of various length, and transaldolase and ....(5) enzymes. The conversion of CO_2 to glucose using the energy from NADPH is similar to that reaction performed in liver (from 3-phosphoglycerate onward) except that NADH is used rather than ....(6).

Answers: (1) 3-phosphoglycerate; (2) six; (3) 12 (4) fructose; (5) transketolase; (6) NADPH.

Complete the following

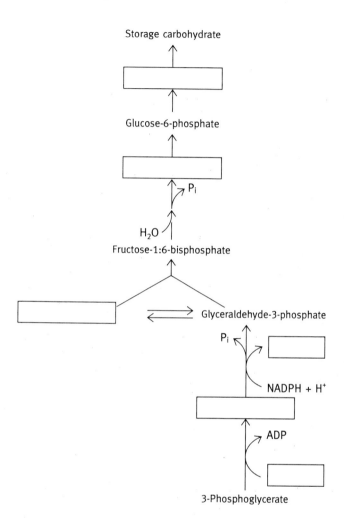

Rubisco

In addition to reacting with CO_2, Rubisco can react with oxygen in a process called
...........................(1), which appears to serve no useful purpose. The
occurrence of these two types of reactions may be related to needs of the plant in
prehistoric times when there was much more ....(2) and much less ....(3) in
the atmosphere. The reasons for the apparent inefficiency of this system are not yet
appreciated. A way to minimize this reaction with oxygen is found in some plants
which utilize ......(4) photosynthesis in which the first product of the synthesis
reaction with CO_2 is the four-carbon molecule ....(5), through the
carboxylation of phosphoenolpyruvate (PEP). Oxaloacetate is reduced to malate

which is then decarboxylated to ....(6). The resulting CO_2 then enters the ....(7) cycle to react with Rubisco, increasing the efficiency of this process by minimizing the energy lost due to the reaction of Rubisco with oxygen. The pyruvate is cycled back through the PEP intermediate. The added energy cost of the C_4 cycle is worth the overall efficiency gained by the availability of higher ....(8) levels for the Calvin cycle.

Answers: (1) photorespiration; (2) CO_2; (3) O_2; (4) C_4; (5) oxaloacetate; (6) pyruvate; (7) Calvin; (8) CO_2.

Review of problems from the end of Chapter 14

- The light reaction requires light, while the dark reaction does not, but can also occur in the light.

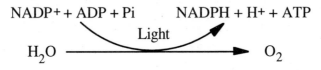

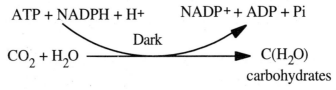

- The chlorophyll molecules that feed excitation energy to the reaction centres are called antenna chlorophylls.
- ATP is generated by the proton gradient that is formed from light energy. If there is no more $NADP^+$ to be reduced, electrons from PSI 'cycle' back to pump more protons and make more ATP.
- $P680^+$ has such a high affinity for electrons that it can take one from H_2O.
- Both processes of proton pumping store the energy from the transfer of electrons and convert it into ATP energy.
- $6C_5 + 6CO_2 \rightarrow 12C_3$ (3-phosphoglycerate).
- CO_2 is converted to sugar in a cyclic process that uses lots of ATP and NADPH. Ribulose-6-phosphate is involved in a catalytic sense, being recycled with every six turns of the cycle. The enzyme ribulose-1:5-bisphosphate carboxylase, 'Rubisco', is of key importance.
- The process of starch synthesis is the same as in gluconeogenesis in liver except that NADPH is used rather than NADH. Glucose is activated as the ADP-adduct for starch synthesis rather than as the UDP-adduct in glycogen formation.

- The wastefulness of the reaction of Rubisco with O_2 can be made somewhat better (ameliorated) by the C_4 plants which generate a CO_2 molecule as a product of one step and thus increase the chance that Rubisco will react with this instead of with an O_2.
- The following is a 'dikinase' reaction that uses the energy from two high-energy phosphate bonds. The enzyme is not found in animals.

$$CH_3-\overset{\overset{\displaystyle O}{\|}}{C}-CO_2^- \quad + \quad ATP \quad \xrightarrow{\quad \overset{\displaystyle Pi}{} \quad} \quad CH_2\!\!=\!\!\overset{\overset{\displaystyle OPO_3^{2-}}{|}}{C}-CO_2^- \quad + AMP + PPi$$

Additional questions for Chapter 14

1. What are the cellular sites of the light-harvesting chlorophyll and of the reactions that convert CO_2 to carbohydrate?
2. What metal ion is in a chlorophyll?
3. What are the reaction centres in photosystem II and photosystem I?
4. Name two molecules involved in the PSII system that contain iron–sulfur centres.
5. How many electrons can be extracted from each water molecule?
6. How does C_3 photosynthesis differ from C_4 photosynthesis?
7. What is the name of the lipid-soluble electron carrier between PSII and the cytochrome *bf* complex?
8. What is the water-splitting centre?
9. Where are the two places to which electrons can be transported from ferredoxin?
10. Which of the photosystems, PSI and PSII, absorbs light?
11. From the reactions of the Calvin cycle, how much energy does it take to make one fructose-1:6-bisphosphate?
12. Where does the energy come from to drive the Calvin cycle?

Chapter 15

....

Amino acid metabolism

Chapter summary

This chapter describes the metabolism of amino acids, focusing on their breakdown for use as energy and highlighting the deamination and transamination reactions involved. The urea cycle and the mechanisms for transporting nitrogen through the blood are of key importance.

Learning objectives

- ❑ The role of proteins in the diet.

- ❑ The distinction between the amino acids that our body can make and those that it cannot.

- ❑ The idea of elemental nitrogen balance.

- ❑ The general fate of amino acids that are digested in excess of the immediate requirements for protein synthesis.

- ❑ The role of the deamination reaction in amino acid degradation and the importance of the reverse of this process in amino acid synthesis

- ❑ The importance of Schiff base formation and breakdown.

- ❑ The importance of the reduction and oxidation reactions that interconvert amino groups and carbon–nitrogen double-bonded compounds.

- ❑ The important role played by glutamic acid in cycling amino groups from other amino acids into ammonia.

- ❑ The importance of the reversibility of the nitrogen cycle in relationship to the synthesis of amino acids.

- ❑ The importance of the amino group transfer cofactor pyridoxal-5′-phosphate (PLP).

- ❑ The distinction between amino acids that are ketogenic and those that are glucogenic.

- ❑ The special reaction that converts phenylalanine to tyrosine catalysed by a monooxygenase.

- ❑ The nature of the abnormality known as phenylketonuria (PKU).

- ❑ The reactions involving S-adenosylmethionine (SAM).

❑ The sequence of steps known as the urea cycle.

❑ The importance of the liver in amino acid metabolism.

❑ The two important pathways for transporting nitrogen through the blood.

❑ The importance of amino acid metabolism under starvation conditions.

A walk through the chapter

Nitrogen balance

Proteins are often a major component of normal diets and, after digestion, the free ....(1) are transported through the ....(2). These amino acids are used for the synthesis of a number of important biomolecules, in addition to other proteins. The uptake involves a selective transport mechanism. Amino acids are not ....(3) for this purpose alone, but ....(4) proteins can be broken down in emergency situations to release their amino acids. Of the 20 common amino acids, the human body cannot make ....(5), and so must depend on their supply from the diet. This can be a problem for certain human populations that have limited diets not containing one or more of these; this results in a condition known as ....(6). Deficiencies of lysine and ....(7) are conditions of special importance. The intake and output of the element nitrogen need to be balanced; thus the fact that proteins are continually being ....(8) must be taken into account. Some of the group of 10 amino acids that our bodies are unable to make are strictly essential, while others can be obtained by conversion re-actions from other ....(9). Amino acids in excess of the immediate requirements for protein synthesis and other specific needs are broken down for ....(10) production or ....(11) storage. Deamination yields keto acids and elemental nitrogen which is incorporated into ....(12). Amino acids can be synthesized by a ....(13) of the deamination. Full discussion of the various pathways that result in the production of the 20 different amino acids in-volves more detail than is needed for the depth of understanding required at this level of biochemistry.

Answers: (1) amino acids; (2) blood; (3) stored; (4) muscle; (5) 10; (6) kwashiorkor; (7) trypto-phan; (8) turned over; (9) amino acids; (10) energy; (11) fuel; (12) urea; (13) reversal.

Complete the following

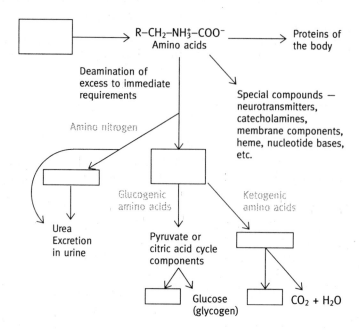

Transamination and deamination

A Schiff base involves the condensation of a carbonyl compound with an
................(1) group. It is ....(2) in the sense that water can be either
the ....(3) of a Schiff base formation or a ....(4) in the
hydrolysis reaction. The carbon–nitrogen double bond can be made into a single
bond via a ....(5) reaction and can thus provide a pathway for
carbonyl-containing keto acids to be made into amino acids. The reverse process
involving an ....(6) can occur as well. The amino acid glutamic acid can
be oxidized and deaminated to give ....(7)   via the enzyme
glutamate ....(8).   This α-keto acid is an intermediate in the
citric acid cycle and can thus be ....(9). Glutamic acid can also give rise
to ....(10) synthesis. Most amino acids couple their deamination to that of
glutamic acid by transferring their nitrogen into this cycle by forming glutamate and
the corresponding keto acid. This represents a two-step process. The first step is a
transamination, forming glutamate and the α-keto acid of the starting amino acid.
The second is the ....(11) of glutamate. Since these reactions are
reversible, amino acids can be ....(12) by a transamination step
between the appropriate α-keto acid and glutamate. The cofactor that participates in
these transaminations is pyridoxal-5′-phosphate (PLP); it acts as an intermediate
accepting the amino group from the donor and handing it to the acceptor. The three-
carbon amino acids, serine and cysteine, can be converted to the three-carbon α-keto

acid, pyruvate. The carbon skeletons of the deaminated amino acids in the form of α-keto acids are termed ketogenic if they can give rise to ....(13) but ....(14) if they can give rise to glucose.

Answers: (1) amino; (2) reversible; (3) product; (4) reactant; (5) reduction; (6) oxidation; (7) α-ketoglutarate; (8) dehydrogenase; (9) oxidized; (10) glucose; (11) deamination; (12) synthesized; (13) acetyl-CoA; (14) glucogenic.

Monooxygenases and SAM

The metabolism of phenylalanine is of special interest as it is converted to ....(1) by an enzyme phenylalanine ....(2), which employs a cofactor molecule called tetrahydrobiopterin (RH_4). This enzyme belongs to the class of enzymes known as the ....(3), or mixed function oxygenases. There is a common abnormality called phenylketonuria (PKU) in which the normal pathway of the conversion of ....(4) to ....(5) is blocked; this results in the buildup of high levels of ....(6), which is found in the urine. This can have severe consequences, especially in children, but can be treated with dietary changes. Other genetic diseases associated with impaired amino acid metabolism are maple syrup disease and alcaptonuria.

The methyl group of ....(7) is important in metabolic reactions involving *S*-adenosylmethionine (SAM); in these transmethylase reactions the ....(8) group is transferred with the aid of ATP hydrolysis energy. The loss of the methyl group by methionine leaves ....(9), which eventually transfers the sulfur to serine forming ....(10). The compounds that gain their methyl group from this SAM mechanism include ....(11), phosphatidylcholine, and ....(12), as well as many of the methyl groups attached to the nucleic acids of such molecules as RNA and ....(13).

Answers: (1) tyrosine; (2) hydroxylase; (3) monooxygenases; (4) phenylalanine; (5) tyrosine; (6) phenylpyruvate; (7) methionine; (8) methyl; (9) homocysteine; (10) cysteine; (11) creatine; (12) epinephrine; (13) DNA.

The urea cycle

The amino nitrogen resulting from the deamination is transferred to the water-......................(1), inert, nontoxic molecule ....(2) in the liver via a reaction involving ....(3); this is known as the urea cycle. Arginine is made from ....(4) by the addition of CO_2 and two nitrogen atoms. Ammonia, CO_2, and a phosphate are combined to form the small molecule ....(5) phosphate with the help of two ATP hydrolysis energy steps. This then combines with ornithine to give ....(6). The final amino group needed to convert citrulline to arginine comes directly from aspartate and involves the comparatively large molecule ....(7).

Answers: (1) soluble; (2) urea; (3) arginine; (4) ornithine; (5) carbamoyl; (6) citrulline; (7) argininosuccinate.

Complete the following

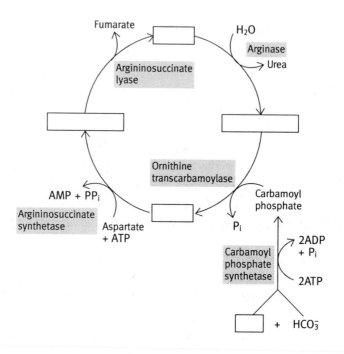

Transport

Urea is formed in the liver and thus the nitrogen atoms from other tissues must be transported to this organ. There are two mechanisms. Firstly, this nitrogen becomes the amide nitrogen of glutamine (not the α-amino group), which is synthesized from glutamate by the enzyme ....(1) synthetase with the help of ATP energy. The resulting glutamine can travel through the blood. Upon reaching the liver, the glutamine can be hydrolysed back to ....(2) with the release of ammonia. Secondly, nitrogen is transported as ....(3), which gains the nitrogen via reaction with the other amino acids by transamination with pyruvate. Once in the liver, the alanine gives up the ammonia and becomes pyruvate which can be used to make ....(4). The entire cycle is called the ....(5) cycle; it plays a role in starvation, when glycogen reserves are exhausted and the liver makes glucose for the brain via protein metabolism.

Answers: (1) glutamine; (2) glutamate; (3) alanine; (4) glucose; (5) glucose–alanine.

Complete the following

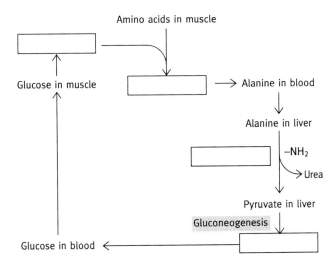

Review of problems from the end of Chapter 15

- In organic chemistry, oxidation reactions often involve adding oxygen atoms. One method of keeping track of when an oxidation reaction has occurred is to count the number of bonds from, for example, carbon to a heteratom other than hydrogen.

$$H-\underset{|}{\overset{|}{C}}-NH_2 \longrightarrow \underset{|}{\overset{|}{C}}=NH \qquad \text{is an oxidation}$$

$$\underset{|}{\overset{|}{C}}=NH \longrightarrow \underset{|}{\overset{|}{C}}=O \qquad \text{is not an oxidation}$$

- Glutamate dehydrogenase deaminates glutamic acid by the mechanism involving an oxidation.
- Transamination for glutamic acid. *Note.* Glutamate and glutamic acid are the same compound without and with a proton, respectively.
- Pyridoxal phosphate carries the NH_2 group from the α-amino position on one amino acid and exchanges it with the oxygen of the carbonyl of α-keto acid. The

original α-keto acid ends up with an α-amino group and the original amino acid ends up as an α-keto acid.

$$R{-}CH{-}CO_2^- \quad + \quad R'{-}\overset{\overset{\textstyle O}{\|}}{C}{-}CO_2^- \quad \longrightarrow \quad R'{-}CH{-}CO_2^- \quad + \quad R{-}\overset{\overset{\textstyle O}{\|}}{C}{-}CO_2^-$$
$$\underset{\textstyle NH_2}{|} \qquad\qquad\qquad\qquad\qquad\qquad\qquad \underset{\textstyle NH_2}{|}$$

It is important to watch for the difference between R and R'.

- Serine and cysteine undergo a special set of deamination reactions, probably because their structures are so similar to that of the very important metabolite, pyruvate. The common intermediate is amino acrylic acid.

- Amino acids are generally converted into citric acid intermediates (glucogenic) or into acetyl-CoA (ketogenic). Only leucine and lysine are purely ketogenic because of the structure of their R-group carbon chains.

- Phenylketonuria involves a defect in the mechanism of conversion of phenyl-alanine to tyrosine which causes the accumulation of phenylpyruvate.

- Tetrahydrobiopterin is a redox cofactor and can be viewed similarly to NAD^+ or FAD:

$$RH_4 \rightarrow RH_2 + 2H^+ + 2e^-.$$

The structural difference between phenylalanine and tyrosine is the —OH group or the aromatic ring.

$$\text{Phe}{-}\bigcirc{-}\text{H} \quad + \quad O_2 \quad \overset{\overset{\textstyle RH_4 \qquad RH_2}{\curvearrowright}}{\longrightarrow} \quad \text{Tyr}{-}\bigcirc{-}\text{OH} \quad + \quad H_2O$$

The enzyme phenylalanine hydroxylase belongs to a class of enzymes known as monooxygenases or, alternatively, mixed function oxygenases.

- S-adenosylmethionine, 'SAM', has a positively charged sulfur with three bonds: one to the methyl group; one to the rest of the methionine; and the third to the 5' carbon of adenosine. This arrangement enables the methyl group to leave easily.

- Make a table to remember these ideas.

Substrate	Structure	Reaction (enzyme)
Arginine	?	Hydrolysis of the Schiff base resonant form releasing urea, H_2NCONH_2 (arginase)
Ornithine	?	Nitrogen is added which has been activated in the form of carbamoyl phosphate (ornithine transcarbamoylase)
Citrulline	?	Is coupled to succinate in a reaction using ATP energy (argininosuccinate)
Argininosuccinate	?	Is cleaved to produce fumarate and arginine so that the cycle can continue (argininosuccinate lyase)

- The need for the urea pathway under conditions of high amino acid intake makes perfect sense, since this is how the system deals with having more amino acids than are needed for protein synthesis. This pathway will also be very active under starvation conditions because amino acids from proteins will be metabolized for energy, and the nitrogen group will have to be removed.
- The urea cycle operates in the liver. Ammonia is attached to glutamate forming glutamine which can travel in the blood. In muscles, the amino acid alanine serves this carrier purpose and is also part of the cycle to bring the energy back to the muscles in the form of glucose.

Additional questions for Chapter 15

1. Name three diseases associated with a malfunction in the normal pathway of amino acid metabolism.
2. How is homocysteine formed and what is the relationship between the structure of this compound and that of the amino acid cysteine?
3. List at least three compounds that can gain their methyl groups from 'SAM'.
4. What amino acid is produced by transamination of oxaloacetate with glutamine?
5. Name two other classes of molecule that can be made from amino acids.
6. What is another name for the cofactor pyridoxal phosphate?
7. How is the pyridoxal phosphate cofactor held in the active site of the enzyme?

8. What class of reaction is catalysed by a dehydratase?

9. In the conversion of homocysteine into cysteine, what other amino acid is used and what is the other product?

10. What coenzyme is involved in the synthesis of glycine as a one-carbon carrier?

11. What is the molecular formula of epinephrine?

12. What is the structure of carbamoylphosphate and why is this compound important?

13. Why not just transport the nitrogen through the blood as ammonia?

14. Why is the choice of alanine as a carrier for blood nitrogen an especially good one?

Chapter 16

··

Garbage disposal units inside cells

Chapter summary

This chapter describes the lysosomes and peroxisomes, which are involved in destroying unwanted molecules and catalysing specialized reactions. The nature of the membrane fusion events involved as well as certain diseases associated with abnormalities in the function of these bioorganelles are discussed.

Learning objectives

- ❑ The primary function of lysosomes.

- ❑ An example of a reaction that occurs in the peroxisome.

- ❑ The mechanism accounting for the way in which lysosomes come in contact with the material to be degraded.

- ❑ The difference between primary and secondary lysosomes.

- ❑ The mechanism for maintaining the unique pH inside the lysosomes.

- ❑ A few diseases associated with abnormalities in the functioning of the lysosomes.

- ❑ The role of hydrogen peroxide in the reactions occurring in peroxisomes.

A walk through the chapter

Lysosomes

Lysosomes are membrane-bound organelles in the cytoplasm of all nucleated cells; their function is the destruction of unwanted molecules and larger structures that find their way into the cell. Peroxisomes are similar but with the special role of dealing with chemical conversions that are difficult for the rest of the cell. Lysosomes contain a mixture of hydrolytic enzymes under acid pH. Maintained by ATP-dependent proton pumps, these enzymes can hydrolyse almost all possible biological molecules.

Complete the following

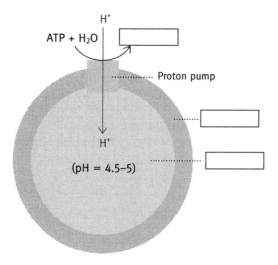

Fusion events

The material that is to be destroyed gets into the lysosome via membrane fusion events. Mitochondria targeted for destruction are engulfed after about 10 days in a membrane vesicle; the resulting vesicle is called an autophagosome. Other molecules are encased in endocytotic vesicles. The lysosome vesicles released by the Golgi fuse with these vesicles and become secondary lysosomes. The recognition that occurs between these vesicles is thought to involve proteins termed 'snares'; v-snares interact with vesicles and t-snares interact with target membranes.

Complete the following

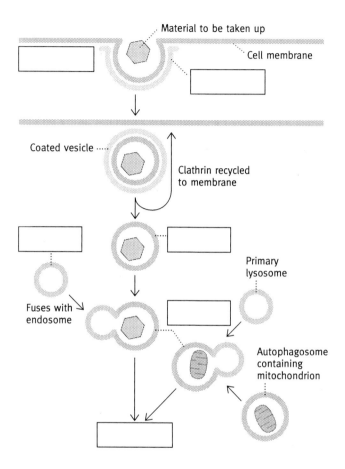

Diseases

The fusion of the vesicles involves ....(1) hydrolysis. It is thought that the digestion products are returned to the ....(2) for reuse. The lysosomes are involved in the constant process of ....(3), which involves many components of the cell. Many diseases can be traced to abnormalities in the functioning of lysosomes, such as ....(4), ...(5) disease, and ....(6) storage disorders. There is still much to be discovered about this class of bioorganelles. ....(7) carry out oxidations producing hydrogen peroxide (H_2O_2), which further ....(8) other species. The list of molecules that undergo reactions involving peroxisomes includes some very long chain ....(9) acids, certain phenols, and the unnatural .....(10) amino acids.

Answers: (1) GTP; (2) cytoplasm; (3) recycling; (4) Tay–Sachs; (5) Pompe's; (6) lysosomal; (7) Peroxisomes; (8) oxidizes; (9) fatty; (10) D-.

Review of problems from the end of Chapter 16

- Lysosomes are involved in the destruction of useless or potentially harmful things.
- The Golgi is a set of flattened membranous bags that sorts out proteins synthesized by the endoplasmic reticulum and sends them to their appropriate destinations.
- Membrane fusion events join the lysosome vesicles with vesicles containing the material to be degraded.
- Major problems occur if the lysosomes do not function properly.
- It is not clear why the reaction that is affected in Pompe's disease is essential for the lysosome.
- Peroxisomes are specialized vesicles that carry out oxidation reactions.

Additional questions for Chapter 16

1. Name three diseases thought to be associated with abnormalities in the functioning of the lysosomes.
2. In which direction are the protons pumped across the membrane of a lysosome?
3. What happens to the clathrin that coats the pit during receptor-mediated endocytosis after the endosome that is formed loses its coat?
4. What happens to the material inside the lysosomes after it has been digested?
5. How long does the average red blood cell circulate in the blood before it is destroyed and what event is thought to trigger its destruction?
6. What would happen to the cell if a lysosome were to rupture?

Chapter 17

··

Enzymic protective mechanisms in the body

Chapter summary

This chapter describes a variety of protective mechanisms involving enzyme reactions. The intrinsic and extrinsic pathways of blood clotting are highlighted and the systems that safeguard against inappropriate clot formation are presented. The mechanisms for the detoxification of xenobiotics are also highlighted with a description of the P450 monooxygenase system.

Learning objectives

- ❏ The primary function of blood clotting.

- ❏ The principle behind a reaction cascade amplification response.

- ❏ The distinction between the intrinsic and extrinsic pathways of activating blood clotting.

- ❏ The importance of the role of thrombin in the formation of fibrin from fibrinogen monomers.

- ❏ The mechanisms safeguarding against inappropriate clot formation and the potentially disastrous results if these systems fail.

- ❏ The roles of vitamin K and Ca^{2+} in the clot formation reaction.

- ❏ The role of protease inhibitors in protecting fragile components of systems from damage by very powerful enzymes like elastase.

- ❏ The reactions that generate reactive oxygen species, such as hydrogen peroxide and the superoxide anion.

- ❏ The nature of chain reactions and the propagation of an unpaired electron.

- ❏ The strategies used by the body to defend against these highly reactive free radicals.

- ❏ The nature of the class of molecules known as the antioxidants.

❑ The general properties of the xenobiotics and the role of oxidation by the cytochrome P450 system.

❑ The reactions of glutathione that involve the formation of a disulfide bond.

❑ The reactions catalysed by the enzymes superoxide dismutase and catalase.

❑ The reactions catalysed by the enzymes glutathione peroxidase and glutathione reductase.

A walk through the chapter

Blood clotting

At the enzymic level the process of blood clotting serves to prevent ....(1). The system is able to respond rapidly by a reaction cascade amplification. The mechanism can be divided into two parts: ....(2) the necessary enzymes; and the actual ....(3) formation reactions. The cascade reaction involves the proteolytic activation of precursor ....(4), which are available and waiting for the signal. The occurrence of a wound causes two things to happen; these are termed the ....(5) and ....(6) pathways. The intrinsic pathway begins when damaged vessels expose structures that initiate blood ....(7) to form a temporary plug. Then a series of blood factors forms a cascade, which eventually activates the protease ....(8), which causes the actual clot to form. The extrinsic pathway involves tissue ....(9) released from damaged cells that also eventually activate thrombin. For normal clotting, both pathways need to be operational. The action of thrombin causes the ....(10) monomer to ....(11) and crosslink to form strands of fibrin which form the basic structure of the clot. Inappropriate clot formation is safeguarded against by a series of ....(12) inhibitors that prevent the clot reaction from spreading beyond the site of the wound. The clot thus formed can also be dissolved by the action of the enzyme ....(13). Inappropriate blood clotting is the underlying cause for a number of ....(14). Warfarin is a chemical rat poison that interferes with the blood clotting process by working as a competitive inhibitor of ....(15) and thus interfering with the necessary role played by .....(16).

Answers: (1) bleeding; (2) activating; (3) clot; (4) proteins; (5) intrinsic; (6) extrinsic; (7) platelets; (8) thrombin; (9) factors; (10) fibrinogen; (11) polymerize; (12) protease; (13) plasmin; (14) deaths; (15) vitamin K; (16) Ca^{2+}.

The detoxification process

Foreign chemicals (known as xenobiotics) include a wide range of chemicals, with many having the common property of being ....(1) soluble rather than water-soluble; they thus accumulate in ....(2) cells and membranes rather than being excreted in the ....(3). These chemicals are treated so as to become more ....(4) by the cytochrome ....(5) system. P450 is a ....(6) protein of the smooth ....(7) reticulum, which ....(8) foreign molecules in a monooxygenase reaction using ....(9) and ....(10). The system is able to attack a large number of compounds. Glucuronidation is another, secondary modification pathway used to deal with foreign molecules. This is a process known as ....(11). Some natural compounds are also decomposed through this P450 mechanism. There is evidence that some compounds that are oxidized by P450 become ....(12) carcinogenic. There is also a system of transporting toxic drugs out of the cell in a process known as multidrug ....(13). The body is protected against its own ....(14). The elastase of ....(15) is prevented from acting at inappropriate places and times by specific ....(16). Maintaining the balance between the amount of protease and that of its inhibitor is critical.

Answers: (1) lipid-; (2) fat; (3) urine; (4) polar; (5) P450; (6) heme; (7) endoplasmic; (8) hydroxylates; (9) O_2; (10) NADPH; (11) detoxification; (12) more; (13) resistance; (14) proteases; (15) neutrophils; (16) inhibitors.

Reactive oxygen species

The molecular gas oxygen, O_2, must accept four ....(1) to reach the reduced form of ....(2). While both O_2 and H_2O are not overly reactive, there are oxygen species in oxidation states between these two that are ....(3) reactive, for example, the superoxide anion which has a ....(4) charge and an ....(5) electron. This molecule is unintentionally formed. Another similarly dangerous species is hydrogen peroxide, ...(6), which is formed by a number of pathways. In addition, some members of this class of molecules are generated initially for specialized purposes. Chain reactions that can result from these radicals can cause a great deal of damage by the propagation of the unpaired ....(7). Three strategies used by the body to defend against these are the vitamins C and ......(8), the superoxide ....(9), and glutathione ....(10) enzyme and other reductase enzyme systems. Vitamin C, which is water-soluble, and vitamin E, which is lipid-soluble, are members of the class of molecules known as ....(11) which can react with the generated free radicals and stop the propagation of the destructive chain reaction. Superoxide dismutase reduces the superoxide radical to ....(12)

which is in turn reduced to water by the enzyme ....(13). Glutathione, the small thiol-containing peptide, functions to keep the thiols of proteins in their ....(14) state, and is also involved as a substrate in a reaction catalysed by glutathione peroxidase that reduces H_2O_2 to H_2O. In this reaction, the glutathione is oxidized to a disulfide and the cycle is completed by the reduction back to glutathione by the enzyme, glutathione ....(15).

Answers: (1) electrons; (2) H_2O; (3) very; (4) negative; (5) unpaired; (6) H_2O_2; (7) electron; (8) E; (9) dismutase; (10) peroxidase; (11) antioxidants; (12) hydrogen peroxide; (13) catalase; (14) reduced; (15) reductase.

Review of problems from the end of Chapter 17

- Cascades are the biological method of amplification.
- Thrombin is a protease that clips off fibrinopeptides from the fibrinogen monomers allowing the monomers to polymerize and form fibrin.
- Enzymically formed covalent crosslinks between amino acid R-groups are involved in the conversion of a soft clot into a more stable structure.
- Vitamin K is a cofactor in the carboxylation of a glutamic acid residue on thrombin which allows the prothrombin to bind Ca^{2+}, a step that is necessary for activation to thrombin.
- The cytochrome P450 catalyses a monooxygenase reaction that uses both O_2 and a reducing agent in the oxidation of a substrate.
- The term 'mixed function oxygenase' is often applied to this class of enzymes because both oxidating and reducing molecules are involved.
- Glucuronyl-UDP is the activated form of glucuronic acid. This can react with nonpolar foreign molecules and make them more polar for disposal in the urine.
- Multidrug resistance is a membrane transport system with a specificity for a wide variety of lipid-soluble compounds.
- Smoking irritates the lungs and initiates a self-destructive reaction involving the enzyme elastase. In addition, chemicals in the smoke inactivate a protease inhibitor, which allows additional damage to occur to the lungs.
- Superoxide is best written $O_2^- \bullet$.
- The dismutase and catalase reactions are of prime importance; in addition, antioxidants can quench the propagation of free radical chain reactions.

Additional questions for Chapter 17

1. What does solubility have to do with the detoxification of the xenobiotics?
2. What effect does warfarin have on the blood clotting process?
3. What happens to large foreign molecules that are ingested?
4. What constitutes a free radical?
5. Where do the intrinsic and extrinsic pathways for blood clotting come together?
6. Which protein factor is missing in patients with hemophilia A?
7. What is the name of the protease that can dissolve blood clots?
8. How does the structure of glucuronide differ from that of glucose?
9. How does ferrihemoglobin differ from normal hemoglobin and what role does glutathione play in this system?
10. What is the connection between the reduction of glutathione, GSSG $\rightarrow$ 2GSH, and the pentose phosphate pathway?

Chapter 18

....

Nucleotide metabolism

Chapter summary

This chapter describes the metabolism (synthesis, degradation, and use) of nucleotides. The structure of this class of compounds is presented as are both the *de novo* and salvage pathways for nucleotide synthesis. The cofactor, tetrahydrofolate, is introduced and a number of diseases and genetic disorders associated with malfunction in these metabolic steps are mentioned.

Learning objectives

❑ The three general structural elements of a nucleotide.

❑ The numbering convention on the ribose sugar with a specific emphasis on the class of 2-deoxyribose nucleotides.

❑ The five aromatic nitrogen bases (A,G,C,U, and T) and the special relationship between U and T.

❑ The classification based on purine and pyrimidine nitrogen bases.

❑ The salvage pathway for purine and the role of the ribotidation reaction.

❑ The importance of the compound 5-phosphoribosyl-1-pyrophosphate (PRPP).

❑ The importance of IMP in the *de novo* synthesis of the purines.

❑ A general understanding of the reactions that involve the cofactor tetrahydrofolate and its formylated derivative.

❑ The pathways for transferring high-energy phosphates between the mono-, di-, and triphosphate nucleotides.

❑ The genetic disorder Lesch–Nyhan syndrome and the disease gout as they relate to nucleotide metabolism.

❑ A general understanding of suicide inhibition and the drug allopurinol.

❑ The types of control mechanisms that regulate the amount of nucleotides produced and also keep the relative amounts balanced.

❑ The ways in which pyrimidine metabolism differs from purine metabolism in terms of the salvage pathway.

❑ The pathway that leads to the deoxyribonucleotides.

❑ The unique reaction, catalysed by thymidylate synthase, that converts dUMP to dTMP.

❑ A general understanding of the mechanism of action of the antifolates that act against leukemia.

❑ The relationship between FH_4 and coenzyme B_{12}.

A walk through the chapter

Structure and nomenclature

Nucleotides have three general structural elements—a phosphate, a ....(1), and a base—while a nucleo*side* lacks the ....(2) group. The sugar component is most often ribose (as in RNA) and 2′-deoxyribose (as in DNA). There are five different bases: adenine, ...(3), cytosine, uracil, and ....(4). Note that U is only found in RNA and T only in ....(5) with the major structural difference between U and T being that T has an extra ....(6) group. The following table gives the names of the bases and their corresponding nucleosides.

Base	Base + sugar = nucleoside
Adenine	Adenosine
Guanine	Guanosine
Cytosine	Cytidine
Uracil	Uridine
Thymine	Thymidine

The bases A and G are ....(7), while the bases C, U, and ...(8) are pyrimidines. Bases are attached to the sugar at the N-9 position of ....(9) and the N-1 position of pyrimidines. The attachment to the sugar involves a β-configuration.

Synthesis

Purines are synthesized *de novo* (from the beginning) with the sugar and phosphate attached to the intermediates as the rest of the molecule is built around this group. There is also a salvage pathway that recovers bases from previous degradation steps. Ribotidation is the reaction which adds the ....(10) and operates the same in both pathways. PRPP (5-phosphoribosyl-1-pyrophosphate) is a sugar with three phosphate groups, two on carbon 1 and one on carbon ....(11); it represents the activated form of ribose-5-phosphate which is thus ready to combine with a base to form a nucleotide, with an ....(12) of the stereochemistry at C-1 of the sugar. In the *de novo* reaction, the nitrogen from ammonia is attached and the rest of the molecule is built around this nitrogen through a nine-reaction sequence that results in IMP. IMP can be converted into either ....(13) or GMP (the two purines). This scheme requires six molecules of ATP. Two of the reactions in this sequence involve one-carbon transfer reactions.

Answers: (1) sugar; (2) phosphate; (3) guanine; (4) thymine; (5) DNA; (6) methyl; (7) purines; (8) T; (9) purines; (10) ribose-5-phosphate; (11) 5; (12) inversion; (13) AMP.

Tetrahydrofolate

The one-carbon group is transferred as a single-carbon aldehyde, the formyl group, from N^{10}-formyltetrahydrofolate. The carrier in this reaction is the coenzyme tetrahydrofolate (FH_4) derived from the vitamin, folic acid (F). This molecule's reactions can be followed using F, FH_2, and FH_4 (and formyl FH_4) with the one-carbon formyl group attached to the nitrogen at the 10 position. The formyl group for this coenzyme comes from the amino acid serine. A one carbon piece (CH_2) is removed from serine leaving glycine, and this methylene carbon makes a new five-membered ring in the cofactor. The CH_2 methylene group is oxidized, using $NADP^+$, to the oxidation state of formaldehyde through an intermediate, meth*enyl* tetrahydrofolate.

Purines and pyrimidines

The purine salvage pathway involves the conversion of free bases to nucleotides by reactions with PRPP. Two phosphoribosyl transferase enzymes are involved, forming AMP, ...(1), or IMP. The salvage pathway is very economical considering the amount of energy required in the *de novo* pathway. The Lesch–Nyhan syndrome results from a ....(2) disorder in which one of the hypoxanthine–guanine phosphoribosyl transferase enzymes (HGPRT) is missing. The disease, *g*out, results from excess production of ....(3) acid due to the lack of a sufficient salvage reaction and the overproduction of purine nucleotides. Purines are thought to travel in the blood from the ....(4) where they are made, but there is much about the salvage pathway that is not completely understood. There are feedback mechanisms to coordinate the efforts of the *de novo* and salvage pathways. Uric acid is the oxidation product obtained from ....(5) and guanine and crystals of this material are associated with the disease ....(6). The action of the drug ....(7) can be understood in terms of its structural similarity to hypoxanthine; it works as an inhibitor of an ....(8) enzyme. This is an example of a ....(9) inhibition, since it is the catalytic action of the enzyme that results in its own inactivation by transforming the drug into alloxanthine, a potent ....(10). The control of purine nucleotide synthesis occurs by ....(11)   feedback control. The first enzyme in the *de novo* path, PRPP synthetase, is negatively modulated by the di- and monophosphate purines. IMP goes to form either AMP or .....(12) and each of these feedback-control their own production. It is important to have a balanced production of ....(13) and GTP.

Answers: (1) GMP; (2) genetic; (3) uric; (4) liver; (5) hypoxanthine; (6) gout; (7) allopurinol; (8) oxidase; (9) suicide; (10) inhibitor; (11) allosteric; (12) GMP; (13) ATP.

Complete the following

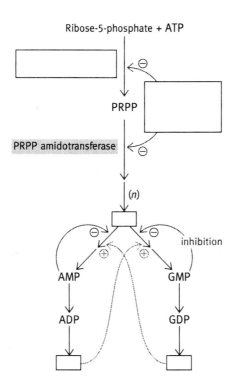

Additional points

Pyrimidine nucleotides are obtained primarily by *de novo* synthesis. Mammals do not appear to have significant salvage pathways in this regard, although there are exceptions. Pyrimidine synthesis starts with ....(1) and contains UDP as an intermediate to forming CTP. The deoxyribonucleotides used in DNA are formed in a reaction that ....(2) the 2'-hydroxyl group to a CH_2 group while the nucleotide is in the diphosphate stage using NADPH. Since dUTP is not used in DNA it is converted to dTTP via dUMP and dTMP. The four deoxynucleotide triphosphates are produced in balanced proportions for use in DNA synthesis. The conversion of dUMP to dTMP is of special interest since it employs the enzyme ....(3) ....(4) and the coenzyme ....(5). In this case the carbon is transferred in the —CH_2— form and becomes the ....(6) group of thymidine monophosphate. Note that some early editions of the main text contain a typographical error in the structure of dTMP on page 233, where one of the ring nitrogens was left out.

The chemistry of this cycle plays an important role in understanding the action of the antileukemia drugs, E methotrexate and ....(7), known as ....(8). These compounds inhibit the FH_2 reductase involved in ....(9) this coenzyme by mimicking the folate structure. Notice the subtle chemical structural difference between the natural compound and these powerful pharmaceuticals. Cancer cells, which are ....(10) growing and thus synthesizing large amounts of DNA, are preferentially inhibited. The FH_4 coenzyme family is also involved in transferring a methyl group from homocysteine to form ....(11). Coenzyme B_{12} is also involved in this cycle and is important in keeping the available supplies of FH_4 from becoming trapped as ....(12) FH_4.

Answers: (1) aspartic acid; (2) reduces; (3) thymidylate; (4) synthase; (5) methylene FH_4; (6) methyl; (7) amethopterin; (8) antifolates; (9) regenerating; (10) rapidly; (11) methionine; (12) methyl-.

Review of problems from the end of Chapter 18

- PRPP (5-phosphoribosyl-1-pyrophosphate) is a ribotidation agent. In this compound, carbon 1 of the sugar is activated with a good leaving group in the form of pyrophosphate. The nitrogen on the aromatic base attacks this carbon in a displacement reaction. The pyrophosphate is hydrolysed to two molecules of inorganic phosphate, driving the reaction in the desired direction.

- The name of the formyl group is derived from formic acid, HCO_2H, a one-carbon compound which is treated in this sense as an aldehyde; thus the formylated-tetrahydrofolate carries the one-carbon aldehyde known as the formyl group.

-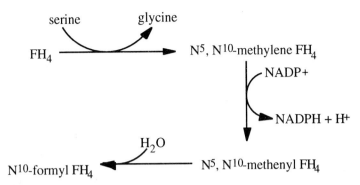

- In the purine salvage pathway the free (without sugars or phosphates) aromatic nitrogen bases, adenine, guanine and hypoxanthine, are converted directly to the monophosphate derivatives. The latter two can both be substrates for the HGPRT (hence the rather cumbersome name). The enzyme that catalyses this step with adenosine could be called adenosine phosphoribosyl transferase (APRT) since a PR group is transferred from PRDP to adenosine.

- Purine salvage reactions do not occur due to the lack of the HGPRT enzyme. Metabolic deficiencies are often more pronounced in children because of their rapid rate of anabolic (synthetic) metabolism.

- Allopurinol causes xanthine oxidase to inhibit itself by suicide inactivation.

-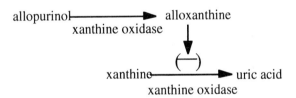

-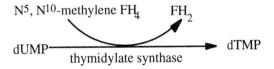

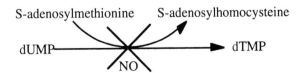

-

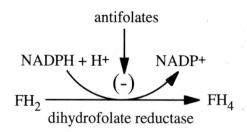

Cancer cells are preferentially affected because of their heightened metabolic rates.

-

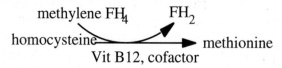

In patients with pernicious anemia it is believed that vitamin B_{12} is absent so that the tetrahydrofolate cofactor is 'trapped' as the methylene derivative.

Additional questions for Chapter 18

1   What does the prime (′) indicate in the names of the nucleotides and why is a prime not needed in the name ribose-5-phosphate?

2.   Which aromatic nitrogen base is found in RNA but not in DNA?

3.   What kind of bond links the phosphate to the 5′ position in monophospho-nucleotides?

4.   Why does it make sense that the stereoconfiguration at C-1 in PRPP has the α configuration?

5.   What is the structural difference between xanthine and hypoxanthine?

6.   What amino acid is part of the structure of folic acid?

7.   What are the two main amino acids used as precursors in the synthesis of IMP?

8.   What is the source of the NH_2 group in GMP?

9.   What kind of reaction is involved in the conversion of N^5, N^{10}-methenyl FH_4 to N^{10}-formyl FH_4?

10.   What are the symptoms and possible treatment for patients lacking adenosine deaminase in lymphocytes?

11.   Explain in general terms the feedback control of the purine nucleotides on their synthesis from ribose-5-phosphate.

12.   What amino acid is at the starting point of the pyrimidine biosynthesis?

13.   What are the two main structural differences between methotrexate and folic acid and which group interferes directly with the chemistry of this cofactor?

Chapter 19

DNA—its structure and arrangement in cells

Chapter summary

This chapter begins the fourth part of the book dealing with information storage and utilization. The chapter describes the all-important molecule DNA, in particular, its structure and arrangement in cells. The structures of the four main bases found in DNA are examined as well as their ability to form hydrogen bonds. The structural difference in the sugar of the RNA is also presented. Current knowledge about the structures of the DNA within the cell is contrasted with some of the questions about this structure that have yet to be answered.

Learning objectives

- ❑ The structural difference between the sugars that make up RNA and DNA.

- ❑ The fact that the aromatic nitrogen bases can exist as tautomeric forms.

- ❑ The differences in polarity between the phosphates, sugars, and aromatic nitrogen bases.

- ❑ The nature of the phosphodiester bond linking the chain of nucleotides together.

- ❑ The meaning of primary nucleic acid structure.

- ❑ Why the linkages between the 2′-deoxy nucleic acids are more stable than those in RNA.

- ❑ The types of intermolecular forces that give stability to the DNA double helix.

- ❑ The Watson–Crick base pairing code A–T and G–C involving one purine and one pyrimidine each.

- ❑ The importance of the processes of melting and annealing in hybridization interactions.

- ❑ Specifics about the structural features of the various double helical forms.

- ❑ The implications of the facts that the two DNA strands have directionality and are antiparallel, having opposite polarity.

❑ The convention of reporting the sequence of bases in a piece of DNA in the 5′ to 3′ direction.

❑ The approximate number of base pairs in human DNA.

❑ The nature of chromatin, nucleosomes, and histones in the structure of DNA packing.

❑ An introductory level understanding of the similarities and differences between eukaryotic and prokaroyotic DNA structure.

❑ A familiarity with some terms associated with the process of eukaryotic cell division as they relate to nuclear structure, including metaphase and interphase chromosomes, a centromere and the kinetochore assembly, and heterochromatin.

❑ The nature of the information contained in the sequence of DNA bases and the definition of a gene.

❑ The occurrence of transposons.

A walk through the chapter

Polynucleotides

Polynucleotides are formed from the ....(1). The phosphates of the DNA are acidic (thus the name) and impart a strong ....(2) charge to the molecule making it very ....(3), although the bases themselves are aromatic and essentially hydrophobic. The linkages that form the backbone of the polymer are ....(4), since two of the oxygens on each phosphate are the oxygens of alcohol substituents. This bridge occurs between the ....(5) of one nucleotide and the 5′-OH of the next. The backbone consists of alternating phosphate–sugar–phosphate groups; the sequence of bases attached to each sugar is known as the ....(6) structure. The uniqueness of the deoxy sugars in DNA as compared to those in RNA stems from the absence of the ....(7) group in the former which makes the DNA structure more chemically stable. The presence of the 2′-OH works as an intramolecular catalyst for the hydrolysis reaction that breaks the polymer of nucleic acids into monomers. The DNA double helix is held together by complementary base pairing known as ....(8) base pairing. In this arrangement, A is across from .....(9) with ....(10) hydrogen bonds and G is across from .....(11) with ....(12) hydrogen bonds. Note that each of the pairs includes one purine and one ....(13).   The process of using heat to separate the two strands is known as melting, with the reverse process called ....(14).   The process by which the correct combinations of bases line up to form complementary interactions is called ....(15).   The most stable three-dimensional structure for the polymers of paired bases is a spiral, or helix with about ....(16) base pairs per rotation. The helices are right-handed, and there are two ....(17): a large one called the major groove and a smaller one called the minor groove. This configuration is known as the ....(18) form. There are other forms of DNA which occur under special circumstances such as the A form and the Z form. The double helix structure needs to be flexible to bend its long length into the relatively small nucleus. The DNA strands are said to be antiparallel because the two strands run in ....(19) directions and have opposite ....(20). There is directionality to the chains of DNA because the ends are different, one with a free 5′ site and the other with a free ....(21) site. By convention, the sequence of each strand is listed in the direction ....(22), and thus, if one strand runs up the page, the other runs down. The 5′ to 3′ refers to the direction traversed across the sugar. *Note*. Some early editions of the main text contain a typographical error on page 244 in which the word 'left' was printed in place of the word 'right'.

Answers: (1) monomers; (2) negative; (3) hydrophilic; (4) phosphodiesters; (5) 3′-OH; (6) primary; (7) 2′-OH; (8) Watson–Crick; (9) T; (10) two; (11) C; (12) three; (13) pyrimidine; (14) annealing; (15) hybridization; (16) 10; (17) grooves; (18) B; (19) opposite; (20) polarities; (21) 3′; (22) 5′ to 3′.

DNA and the nucleus

Individual DNA molecules can be enormous. Human DNA contains ....(1) base pairs. The packaging of DNA into the nucleus is not completely understood. The total length of DNA in a human cell is about ....(2). The nucleus has a diameter about 1 million times smaller. DNA in eukaryotic cells exists as a complex with proteins called ....(3); the main proteins are ....(4) and contain a great number of ....(5) charged amino acids. There is a ....(6) core made of a complex of eight protein molecules on to which the DNA wraps in two turns. The nucleosomes are separated on the strands by about ....(7) pairs, and the strings of nucleosomes are arranged to form a fibre which then forms loops that are attached to a central chromosomal protein ....(8). These loops are further packed into structures which have not yet been determined. During the process of eukaryotic cell division, the DNA appears in compact structures called ....(9) chromosomes. These structures contain a ....(10) section of DNA in the middle to which the ....(11) assembly of proteins is attached. After cell division the chromosome becomes unpacked into the ....(12) chromosome, although some regions of the DNA remain in a very condensed structure called ....(13). The less condensed functional regions are called ....(14). The degree of packing of the DNA in chromatin is relevant to gene expression.

Answers: (1) 6 billion; (2) 1–2 metres; (3) chromatin; (4) histones; (5) positively; (6) nucleosome; (7) 30–40 base; (8) scaffolding; (9) metaphase; (10) centromere; (11) kinetochore; (12) interphase; (13) heterochromatin; (14) euchromatin.

Complete the following

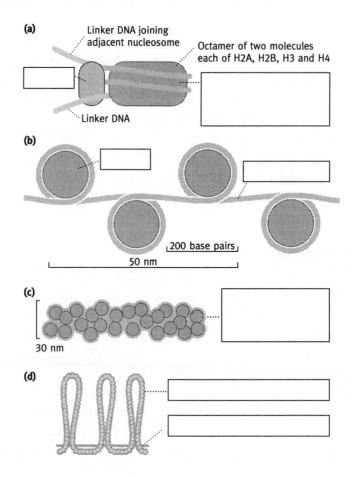

Genes and other DNA elements

The ....(1) of bases on the DNA contains the information for the sequence of ....(2) in proteins. Each stretch of DNA that carries the information for a polypeptide is called a ....(3). There can be multiple copies of the same gene and genes can ....(4) for things other than proteins. Chromosomes can also have ....(5) that are mobile and move from one place in the DNA to another; these elements are called ....(6). Not all of a cell's DNA is in the form of genes as there are sections called ....(7)

sequences for which no established function is known. There is also satellite DNA, which is part of the ....(8) structure and gives very characteristic patterns when analysed due to its ....(9)   and short segments of DNA repeats that run in a tandem arrangement.

Answers: (1) sequence; (2) amino acids; (3) gene; (4) encode; (5) elements; (6) transposons; (7) Alu; (8) heterochromatin; (9) gigantic size.

Review of problems from the end of Chapter 19

- The important aspect of the linkage between nucleic acids is what is happening at the sugar. All of the sugars used in RNA and DNA have both a 3′ and a 5′ OH group. The oxygens of these groups become attached directly to the phosphorus when the phosphate diester bond is formed. Note that when sugars are drawn certain liberties are often taken, such as the fact that bonds without groups on the end represent hydrogens; under more rigorous organic chemical discussions, bonds without a specific group designated on the end would represent methyl groups.

- DNA is chemically more stable than RNA.

- The helical structure minimizes the exposure of the hydrophobic bases to the polar solvent water.

- The right-handed helix is called B DNA and has 10 base pairs/turn.

- The strands of DNA in a double helix each have directionality and they run in opposite directions.

- In simplest terms, the numbering 5′→3′ refers to the position of the hydroxyl on the free end of the chain. The convention is to list the sequence from 5′ to 3′. Note that, because of the stereochemistry of the sugar, in order to show the complete structure one of the strands is written upside down on the paper.

- The base pairing is G with C and A with T. Read the pairs from the 5′ to the 3′ end.

- Nucleosomes are histone–DNA complexes which are a protection against reaction with DNase enzymes.

- Alu sequences are repeated sequences of bases found many times throughout the chromosome.

Additional questions for Chapter 19

1. In the sequence 5′ AGTC 3′ which hydroxyl on the G is linked with the phosphate to the A?

2. What is a major piece of evidence that we don't yet understand about the structure of DNA?

3. When a 5′ phosphate on one nucleotide forms a phosphoester bond with a 3′ hydroxyl from a second nucleotide, what is the other product?

4. What is the structure that is formed when the 2′ OH of a nucleic acid is involved with breaking the phosphate linkage between two bases?

5. How many H-bonds between A–T and G–C?

6. How many base pairs of DNA link together the nucleosomes?

7. Approximately how many base pairs are there in human DNA?

8. If human DNA from a single cell were stretched out straight, how long would it be?

9. What does the sequence of bases of DNA mean?

10. What is the complementary sequence to 5′ GGATTCCATGC 3′?

Chapter 20

....

DNA synthesis and repair

Chapter summary

This chapter describes the cellular reactions that synthesize and repair DNA. The control of initiation of DNA replication in *E. coli* is highlighted and compared with the system in eukaryotes. The problems associated with overwinding and underwinding the DNA helix are addressed as are the enzyme systems that control these events. The polymerase reaction that adds bases to the growing chain is presented along with the replicative fork that allows both strands to be copied simultaneously. The mechanisms for repairing DNA in the pro- and eukaryotic systems are compared. The interesting implications of telomeric DNA are presented along with the occurrence of transposons.

Learning objectives

- [ ] An understanding of some fundamental differences between the bacterial *E.coli* system and the eukaryotic cell system for DNA synthesis.

- [ ] The semiconservative nature of the replication of DNA.

- [ ] That eukaryotic replication has hundreds of origins of replication and that the replication proceeds in both directions.

- [ ] The role of the proteins DnaA and DnaB (also known as helicase).

- [ ] The relationship between (+) and (−) supercoils of DNA.

- [ ] The mechanism for relieving the (+) supercoils formed when the DNA is unwound by the enzymes topoisomerase types I and II.

- [ ] The relationship between strand coiling and winding around the nucleosomes.

- [ ] The role of the single-strand binding protein (SSB).

- [ ] The role of the enzymes, DNA polymerases (Pol I, II, and III).

- [ ] The role of the DNA template and primer.

❑ That bases are added to the 3′ end of the primer so that the new chain grows in the 5′ to 3′ direction.

❑ The role of RNA in the initiation of the synthesis of the new chain.

❑ The major differences between RNA polymerase and DNA polymerase.

❑ The relationship between the directionalities of the two polymerases which are linked together and replicating the two strands of a piece of DNA.

❑ The relationship between the leading and lagging DNA strands and the movement of the loop system.

❑ The occurrence of Okazaki fragments on the lagging strand.

❑ The role of the exonuclease activity on DNA polymerase I.

❑ The nature of proofreading.

❑ The difference between high and low processivity.

❑ The role of DNA ligase.

❑ The role of the proteins Mut S, L, and H, and the reason for methylating the adenine of each GATC sequence on the parent strand.

❑ Some examples of chemical damage that can occur to DNA.

❑ Three different mechanisms of DNA repair.

❑ The problem of deamination and the fact that a deaminated C is the same as a U.

❑ The role of the sliding clamp mechanism also known as the proliferating cell nuclear antigen (PCNA).

❑ The fact that, since eukaryotic DNA is linear, the 3′ ends are not fully replicated by the lagging strand and the role of the telomerase enzyme.

A walk through the chapter

Replication

Each time a cell divides, the chromosomes containing the DNA must be ....(1). The replication of DNA is described as ....(2), since each of the two parent strands acts as a ....(3) for making a new strand. Strand separation is essential for the process of replication. In *E. coli* where the DNA is in a complete ....(4), two replication ....(5) are formed at the origin of replication and move in opposite directions. Synthesis occurs at a rate of about ....(6) base pair copies per second. After the new strands have been added, the interlocked circles are separated by the ....(7) type II enzyme. In eukaryotic systems the rate of replication is ....(8) and, to keep up with the demand, hundreds of ....(9) of replication are initiated and the ....(10) proceeds in both directions. Each stretch of DNA whose replication is under the control of a single origin of replication is called a ....(11). The origin of replication has a specific base sequence unusually rich in A–T pairs to facilitate strand separation, since ....(12) pairs have only two H-bonds and are not as strongly held together as G–C pairs which have three. The binding of multiple copies of the protein ....(13) causes strand separation. The main unwinding protein is known as a ....(14), in this case called DnaB. The cell cycle of eukaryotic cells is more complex than that of bacteria with ....(15) distinct phases; DNA synthesis occurs during the .....(16) phase. Mammalian cells require a mitogenic signal from outside the cell, often in the form of a protein growth factor.

Answers: (1) replicated; (2) semiconservative; (3) template; (4) circle; (5) forks; (6) 500; (7) topoisomerase; (8) slower; (9) origins; (10) replication; (11) replicon; (12) A–T (13) DnaA; (14) helicase; (15) four; (16) S.

Supercoiling

The DNA strands experience supercoiling due to strain on the helix from either being ....(1) (a (+) supercoil) or underwound (a (–) supercoil). This occurs because separation of the DNA strands demands that the duplex ....(2) causing overwinding. The process of relieving the positive supercoils ....(3) of the replicative fork necessarily involves the transient ....(4) of the polynucleotide chain. Enzymes of the class called ....(5), types I and II, catalyse this process. The type I enzyme ....(6) one of the strands permitting the duplex on one side of the break to rotate with respect to the duplex on the other side and release the strain of supercoiling, followed by the resealing of the break via a mechanism that involves a covalent bond between the DNA and one of the protein ....(7) side chains. In *E. coli* the topoisomerase type II, also called ....(8), handles the problem of supercoiling DNA by inserting ....(9) supercoils, which neutralize the positive supercoils. The mechanism involves breaking two strands of the DNA and then transferring the DNA duplex of a coil through the double gap by a mechanism in which one strand of the DNA is passed ....(10) of the other in a step that requires ....(11). The insertion of a negative supercoil is equivalent to the ....(12) of a positive supercoil.

	Effect on DNA	
System	Type I	Type II
E. coli	...............................(13)	...............................(14)
Eukaryote	...............................(15)	...............................(16)

The balance between the activities of gyrase and topoisomerase I in *E. coli* maintains the correct degree of supercoiling. Eukaryotic DNA is also found in the ....(17) state, but the mechanism for generating this does not involve a topoisomerase; rather it is related to the way in which the DNA is wound around the ....(18) in the chromatin assembly by a mechanism in which local positive supercoiling is relaxed, thus introducing a net increase in the amount of negative supercoiling.

Answers: (1) overwound; (2) rotates; (3) ahead; (4) breakage; (5) topoisomerases; (6) breaks; (7) tyrosine; (8) gyrase; (9) negative; (10) over the top; (11) ATP; (12) relaxation; (13) Relaxes (–) supercoils; (14) Inserts (–) supercoils; (15) Relaxes both (+) and (–) supercoils; (16) Relaxes (+) supercoils; (17) underwound; (18) nucleosomes.

Initiation of replication

The enzymes known as DNA ....(1) (Pol I, II, and III) catalyse the reaction of elongating a DNA strand using the deoxyribonucleoside ....(2) (dATP, dCTP, dGTP, and dTTP) as substrates. The reaction requires a piece of DNA to copy called the ....(3). The piece that is synthesized is complementary to this template. A pre-existing piece of DNA to which the growing chain can be attached is needed; this is called a ....(4). The primer grows in length as individual bases are attached to the .....(5) end; thus the chain grows in the 5′ to 3′ direction. This is the reason for the convention of referring to the DNA sequence beginning with the base at the 5′ end and reporting the sequence in the ....(6) direction. RNA is involved in the initiation of the synthesis of the new chain. Recall that the two major differences between RNA and DNA are the ....(7) group on 2′ carbon and the use of U instead of ....(8) in RNA. The enzyme RNA polymerase is different from DNA polymerase in that it *can* ....(9) new chains so that only a ....(10) (along with the necessary nucleotide triphosphates) is required. The enzyme that performs this function is known as ....(11) and, once a small piece of ....(12) has been made, the DNA polymerase can start the elongation reaction. The replicative ....(13) separate the two strands so that two DNA polymerase enzymes can progress along the chromosome. The single-strand binding protein SSB can bind anywhere along the single strand and this helps to drive the strand separation process and prevents the single-stranded form from ....(14). These two polymerases are linked together and must therefore move in the same direction. In order for synthesis to occur at the new strand in the 5′ to 3′ direction, the ....(15) must be read in the 3′ to 5′ direction thus making the new strand ....(16) and forming a duplex. This means that, as the polymerase complex moves across the two template strands, only one of the ....(17) is moving in the correct direction with respect to the polarity of one of the DNA strands. The DNA strand that is in the correct polarity is called the ....(18) strand. The other strand, which is not being replicated in a continuous manner as the polymerase complex moves, is called the ....(19) strand. Replication of the lagging strand is initiated by the action of a primase which makes a number of short primer strands that are separated by short distances of DNA. The DNA that is elongated to these primers on the lagging strand is referred to as Okazaki fragments.

Answers: (1) polymerases; (2) triphosphates; (3) template; (4) primer; (5) 3′; (6) 5′ to 3′; (7) OH; (8) T; (9) initiate; (10) template; (11) primase; (12) RNA; (13) forks; (14) reannealing; (15) template; (16) antiparallel; (17) polymerases; (18) leading; (19) lagging.

Complete the following

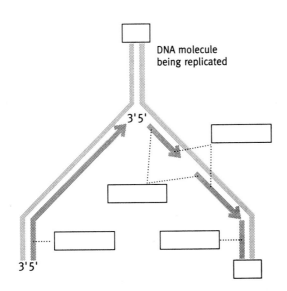

DNA molecule being replicated

3'5'

3'5'

3'5'

The polymerase reaction

In *E. coli* the enzyme polymerase III is held on to the ....(1) by a clamp-type structure of protein subunits with a hole just big enough for the DNA molecule to pass through. The symmetry of this design can be appreciated. As the ....(2) unwinds the DNA double helix and the ....(3) lays down short pieces of RNA separated by short segments of open DNA, the ....(4) template strand is looped back for a short distance putting that part of the DNA into the correct polarity (3′ to 5′ for a template), thus allowing DNA elongation to occur ....(5) at both strands simultaneously. The loop system moves along with the polymerase enzymes while each Okazaki fragment is synthesized; then a new loop is formed. After the polymerase system moves by, the Okazaki fragments are annealed in line to the template DNA but separated by a ....(6)

at the DNA/RNA junction. DNA polymerase I (Pol I) converts these separate pieces into a continuous strand of DNA by ....(7)   the RNA, ...(8) it with DNA, and ....(9) the pieces together. The order of these events is as follows. First, the DNA of the previous Okazaki fragment is elongated to overlap with the RNA template of the following fragment. This addition of DNA occurs in the normal 5′ to 3′ direction. The RNA bases that served as the primer are chopped out one by one from the 5′ end of the primer (5′–3′ ....(10)   activity) as the Pol I enzyme moves down the DNA elongating the previous Okazaki fragment. This process continues until there are no more ....(11) bases between the pieces of newly synthesized DNA. The polymerase I enzyme also has the ability to ....(12) the newly synthesized DNA; it does this by cutting out individual bases that are not correctly base paired with the template strand. The Pol I enzyme only stays firmly attached to the DNA strand long enough to complete the replacement of the RNA primary. It is said to have low ....(13). The nick that is left between what were the Okazaki fragments is closed by the enzyme called ....(14).   The sequence of steps by which this happens involves intermediates in which the enzyme has a covalently bound AMP that is transferred to the 5′ end of the DNA, thus ....(15) this group so that the 3′ hydroxyl of the adjacent DNA base can displace the AMP and form the new ....(16) bond. ....(17) of the pyrophosphate PP_i released in the first step energetically drives the reaction.

Mistakes in the sequence of bases in the newly synthesized DNA are ....(18)   and occur at the rate of about one in 10^5–10^6 base pairs. The polymerase III enzyme also performs a proofreading function and the final error rate is on the order of less than one mistake for every ....(19)   base pairs. A second mechanism to ensure accurate replication of the DNA in E. coli involves three proteins named Mut S, ...(20),   and Mut H. The process involves identifying a mismatch and cutting out and replacing the error on the newly synthesized strand. The bases of the newly synthesized strand are distinguished from the template by the fact that each ....(21)   sequence on the parent (or template) strand has a ....(22) group covalently incorporated to the ....(23)   of this sequence. Thus the strand ....(24) these methylated adenines is the newly synthesized one. By this mechanism the newly synthesized DNA is marked with a ....(25) adjacent to the GATC without the ....(26) group, and all of the DNA bases between this nick and the mismatched pair are replaced.

Answers: (1) template; (2) helicase; (3) primase; (4) lagging; (5) 5′ to 3′; (6) nick; (7) removing; (8) replacing; (9) joining; (10) exonuclease; (11) RNA; (12) proofread; (13) processivity; (14) DNA ligase; (15) activating; (16) phosphodiester; (17) Hydrolysis; (18) inevitable; (19) 10^{+10}; (20) Mut L; (21) GATC; (22) methyl; (23) adenine; (24) without; (25) nick; (26) methyl.

Complete the following

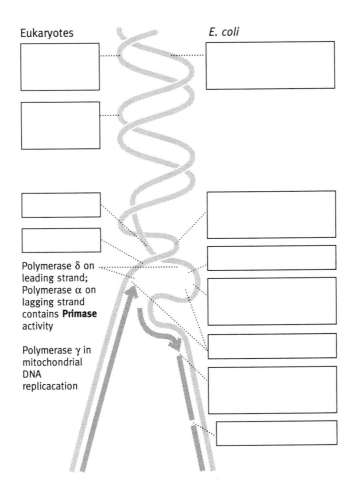

Eukaryotes   *E. coli*

Polymerase δ on
leading strand;
Polymerase α on
lagging strand
contains **Primase**
activity

Polymerase γ in
mitochondrial
DNA
replicacation

Repair of DNA

DNA is also susceptible to ....(1) change or ....(2) over time after it has been synthesized. Reactions, such as the loss of a base (known as ....(3) and ....(4)) and the deamination of cytosine and adenine, occur in a ....(5) but spontaneous and random fashion. Damage to the DNA can also occur due to ....(6) radiation (cf. radioactivity), oxygen free ....(7), carcinogens, and ......(8) light. When other molecules such as proteins become damaged they are ....(9), but DNA must be ....(10). It is often the case that only one strand of the duplex is affected and the other can be readily repaired. There are a number of different types of repair mechanisms including ....(11) repair which operates for example when thymine ....(12) are formed from exposure to UV light and the abnormal bonds are removed and the original base structures restored. Nucleotide ....(13) repair involves the removal of DNA in the region of a lesion or abnormality that ....(14) the double helix. This process involves ....(15) activity which removes bases from the faulty strand followed by a polymerase and ....(16) that complete the process. In base ....(17) repair and AP site repair only the base portion of the nucleotide is removed leaving an AP (apurine or apyrimidine) phosphate sugar still ....(18) to the strand. Repair of an AP site involves nicking, removing, replacing, and resealing the DNA chain. Deamination, which occurs naturally at a small but finite rate, converts ....(19) to uracil and adenine to ....(20). These abnormal bases then need to be removed. It is interesting to note that, if DNA contained U instead of T as it does, then it would be impossible to distinguish between a ....(21) U and a ....(22) C. The base U does not create as much of a problem in RNA because this polynucleotide serves a purpose for a much ....(23) time span than does DNA and thus ....(24) is not repaired.

Answers: (1) chemical; (2) damage; (3) depurination; (4) depyrimidation; (5) slow; (6) ionizing; (7) radicals; (8) UV; (9) destroyed; (10) repaired; (11) direct; (12) dimers; (13) excision; (14) distorts; (15) endonuclease; (16) ligase; (17) excision; (18) attached; (19) cytosine; (20) hypoxanthine; (21) normal; (22) deaminated; (23) shorter; (24) RNA.

Additional topics

The eukaryotic system has many similarities to and a few differences with the *E. coli* system. The eukaryotic system has ....(1) DNA polymerases named with Greek letters. The sliding-clamp mechanism is called ....(2) (PCNA). Eukaryotic DNA is ....(3) while the prokaryotic ....(4) is ....(5). Because of this difference, the ....(6) ends of these linear pieces of DNA are ....(7) fully replicated by the ....(8) strand synthesis mechanism, since replication of the very 3′ end would involve an ....(9) fragment. And, since the ....(10) for an Okazaki fragment is ....(11) which is removed, the very last (or first) few bases on the 3′ end of the parent are not ....(12). The DNA segment on the 3′ end of the parent is called ....(13) DNA and contains no information; in humans it consists of hundreds of repeats of the sequence TTAGGG. This telomeric DNA can be ....(14) to the parent DNA as needed in rapidly ....(15) cells. The enzyme that adds these telomeric sequences is called ....(16) and has the unique characteristics of carrying its own ....(17) in its structure and acting as a ....(18) transcriptase, making DNA from an ....(19) template. Interestingly, the enzyme is found in ....(20) cells, which are continuously rapidly dividing, but not in ....(21) cells implying (speculatively) that somatic cells receive their initial 'ration' of telomeric DNA to suffice for the ....(22) of that cell and its progeny. It has been observed that some ....(23) can 'jump' or otherwise move from one place in the chromosome to another place in the same or a different chromosome. These are called ....(24) and this event is catalysed by an enzyme called a ....(25). A gene of this type called IS, ...(26), has been found in *E. coli*. Interestingly, this gene encodes for the transposase ....(27). There are variations on this model in which the transposon is replicated and the copy inserted elsewhere, a process that could create multiple identical sequences in the genome leading to ....(28) recombination.

Answers: (1) five; (2) proliferating cell nuclear antigen; (3) linear; (4) DNA; (5) circular; (6) 3'; (7) not; (8) lagging; (9) Okazaki; (10) primer; (11) RNA; (12) replicated; (13) telomeric; (14) added; (15) dividing; (16) telomerase; (17) template; (18) reverse; (19) RNA; (20) germ; (21) somatic; (22) lifetime; (23) genes; (24) transposons; (25) transposase; (26) insertion sequence; (27) enzyme; (28) homologous.

Review of problems from the end of Chapter 20

- Replicon—a region of the DNA that is all copied from a single starting point.
- Topological questions deal with the overall three-dimensional shape and relative configurations.
- The relaxation of (+) supercoils and the introduction of (−) supercoils have equivalent effects.
- The main difference between topoisomerase I and II is that type I breaks only one strand of the DNA while type II breaks both strands.
- The action of winding the DNA around the nucleosome adds negative supercoils.
- All four deoxynucleotide triphosphates are needed for DNA synthesis.
- A–U pairing is the same as A–T pairing with U occurring in RNA and T occurring in DNA. The structural difference between U and T is the occurrence of a methyl group in T. Along with this similarity, cytosine differs from uracil in that C has an amino and U has an oxygen in the same position. The chemical reaction in which the amino group of cytosine is lost, leaving a uracil base, occurs at a slow but constant rate; since uracil is not a normal component of DNA, the occurrence of this base is indicative of the deamination of a cytosine. In RNA, deaminated cytosines cannot be detected since uracil occurs naturally on its own.
- A primer piece is needed for the new DNA bases to be added. The new bases are added to the free 3'-OH of the terminal ribose sugar; thus growth of the chain proceeds in the 5' to 3' direction.

- The bases to be added start as free molecules with three phosphate linkages. In the final state the base is paired into the helix and these phosphate bonds have been either broken or changed into lower-energy compounds with the release of free energy.

- Processivity is a measure of how long the polymerase enzyme stays attached to the DNA template and replicates the DNA. This is achieved with a 'ring clamp' structure.

- Polymerase I has three separate enzymatic activities. Pol. I attaches to the nicks between the successive Okazaki fragments and adds nucleotides. It also chops bases off what was the RNA template of the next Okazaki fragment until the newly synthesized piece overlaps with the previously synthesized piece. Finally, this exonuclease activity can also serve the purpose of proofreading and removing mismatched base pairs.

- The correct base pairing of the incoming nucleotide triphosphate with the template dictates which base should be added to the chain. When a mistake is made, the mismatch of H-bonds is apparent and a proofreading mechanism can break the covalent phosphodiester bonds that hold the wrong base attached to the growing chain.

- An associated problem with the occurrence of a mismatch is to determine which of the two mismatched bases is the incorrect one to be removed and which is the correct one which should be used as the template in the process of repairing the mistake. The mechanism is best understood in *E. coli* and is called the methyl-directed pathway for mismatch repair; it is based on the activity of an enzyme that methylates the adenines of GATC sequences in notice DNA. Since the proofreading of newly synthesized DNA occurs before the methylations can take place, the new strand, which is the one that would have an incorrect base sequence, does not have methylated adenines.

- UV irradiation can cause two adjacent thymine bases to form a covalent dimer. The mutation can be repaired by a specific reaction that undoes the dimerization, or the two bases can be removed and replaced.

- The 3′ end of the lagging strand contains a piece of RNA used as the primer. This RNA does not stay permanently attached so the new piece of DNA is then that much shorter.

- Telomeric DNA is added so that the bases that are lost because the 3′ end of the lagging strand does not get copied are bases that are not needed because their sequence does not carry information.

Additional questions for Chapter 20

1. How fast can DNA be made by *E. coli*?
2. What is the role of the topoisomerase type II enzyme in *E. coli*?
3. What are the frequencies of mistakes in the sequence of newly synthesized DNA and that of the final DNA product? What is responsible for this difference?
4. What role does RNA play in the replication of DNA?
5. What is the difference between the leading and lagging strands of DNA?
6. What is the role of the DNA ligase enzyme?
7. What does PCNA stand for and what is it?
8. Which specific atoms are involved in the bond made by DNA polymerase?
9. What is another name for DnaA?
10. Deaminated cytosine (with an oxygen added) is the same as what base?
11. What are three mechanisms of DNA repair?

Chapter 21

....

Gene transcription—the first step in the mechanism by which genes direct protein synthesis

Chapter summary

This chapter describes gene transcription with an emphasis on the mechanisms of controlling gene expression. The differences between the synthesis of mRNA and that of DNA are compared and contrasted. The various promoter boxes and transcription factors functioning in the prokaryotic and eukaryotic systems are presented. Some of the specific details associated with the *lac* and *trp* operons are highlighted. The relationship between introns and exons is important for understanding how the information needed to synthesize a protein is conveyed. The general structural types of the DNA binding proteins including zinc fingers and leucine zippers are also described.

Learning objectives

- ❑ The role of mRNA in carrying the genetic code from the nucleus to the cytoplasm of the cell.

- ❑ The three main structural differences between RNA and DNA.

- ❑ The similarities and differences between the polymerase reaction that makes RNA for transcription and the DNA replication reaction.

- ❑ The size and half-life of RNA and how these compare to those of DNA.

- ❑ The complementary nature of the mRNA, which is made to the template strand, and the convention of referring to the nontemplate DNA strand as the coding or sense strand.

- ❑ The role of the mRNA untranslated regions (UTRs).

- ❑ The relationship between upstream and downstream nucleotides of DNA and the numbering convention of the bases with respect to the initiation site for transcription.

❑ The fact that the roles of template and nontemplate for the DNA strands can be different for different strands.

❑ The three phases of gene transcription.

❑ The occurrence of the Pribnow box (TATAAT) in *E. coli* and the role of the sigma factor protein.

❑ The nature of the hairpin structure causing an RNA self-duplex to form and its relationship to the termination signal.

❑ The nature of the Rho factor in the mechanism of transcription termination.

❑ The nature of genes that are constitutively expressed.

❑ The relationship between strong and weak promoters and the factors that influence these parameters.

❑ The enzymes that would be specifically induced in *E. coli* in response to conditions where lactose is the only source of carbon.

❑ The relationship between genes in the same operon and the occurrence of polycistronic mRNA.

❑ The role of the CAP and the mechanism of controlling the *lac* operon including the role of cAMP and metabolites of lactose.

❑ The mechanism of attenuation of the *trp* operon.

❑ The way in which the control of eukaryotic gene expression differs from that in prokaryotes.

❑ The role of the RNA capping reaction with methylation at the N-7 position of the GTP cap.

❑ The occurrence of introns and exons in the mRNA strand and the process of splicing the mRNA to put the exon pieces together catalysed by spliceosomes.

❑ The importance of the intron-splicing reaction in *Tetrahymena*, the first discovery of a reaction catalysed by an RNA molecule instead of by a protein enzyme.

❑ The relationship between the occurrence of introns and protein domains.

❑ The differences between the 'introns late' and 'introns early' models to explain the occurrence of split genes.

❑ The importance of TATA boxes, enhancer elements, and protein transcription factors in eukaryotic gene expression.

❑ The importance of TATA binding protein (TBP) and TBP-associated factors (TAFS) in gene regulation.

❑ The factors that allow the mechanism for eukaryotic gene control to be so flexible.

❑ The general types of DNA binding proteins based on structural similarities.

❑ The occurrence of palindromic sequences.

❑ The aspect of the leucine zipper proteins that is a misnomer.

❏ The nature of homeodomain proteins and the occurrence of the homeobox.

❏ The relationship between the stability of mRNA and the rate of protein synthesis and the influence of this on the regulation of gene expression.

❏ The role of the iron-responsive elements (IRE) in regulating the synthesis of transferrin.

❏ The occurrence of adenine/uracil-rich elements (AURE) and their effect on mRNA stability.

❏ The fact that mitochondria do not have all of the DNA needed to completely reproduce.

❏ The occurrence of mRNA editing mechanisms.

A walk through the chapter

Messenger RNA

In ....(1)   the DNA is enclosed inside the ....(2), while the protein-synthesizing machinery is in the ....(3). Copies of the coded information are sent out in the form of messenger RNA (mRNA). There are ....(4) main differences between mRNA and DNA: the sugar is ....(5), not deoxyribose, and thus has an OH in the 2′ position; mRNA is ....(6) stranded; and U is used to pair with A instead of T. The enzyme RNA ....(7) catalyses the formation of RNA, using ....(8) as a template, and making RNA in the ....(9)   direction. The antiparallel template is thus read in the 3′ to 5′ direction. The mechanism of each base addition involves the appropriate nucleotide triphosphate with reaction between the ....(10) of the ....(11) chain and the ....(12) group closest to the sugar on the nucleotide triphosphate. The hydrolysis of the pyrophosphate formed energetically drives the reaction. Unlike the DNA polymerase system, the RNA polymerase does not need a ....(13).   Each mRNA encodes for a single gene (or a small group). mRNAs are rather short in length and have a half-life of about 20 minutes to several hours in ....(14) and about 2 minutes in bacteria. A continuous stream of mRNA is thus needed to support the synthesis of a protein. ....(15) mRNA molecules found in bacteria carry the coded instructions for several proteins.

Answers: (1) eukaryotes; (2) nucleus; (3) cytoplasm; (4) three; (5) ribose; (6) single-; (7) polymerase; (8) DNA; (9) 5′ to 3′; (10) 3′-OH; (11) growing; (12) phosphate; (13) primer; (14) eukaryotes; (15) Polycistronic.

Complete the following

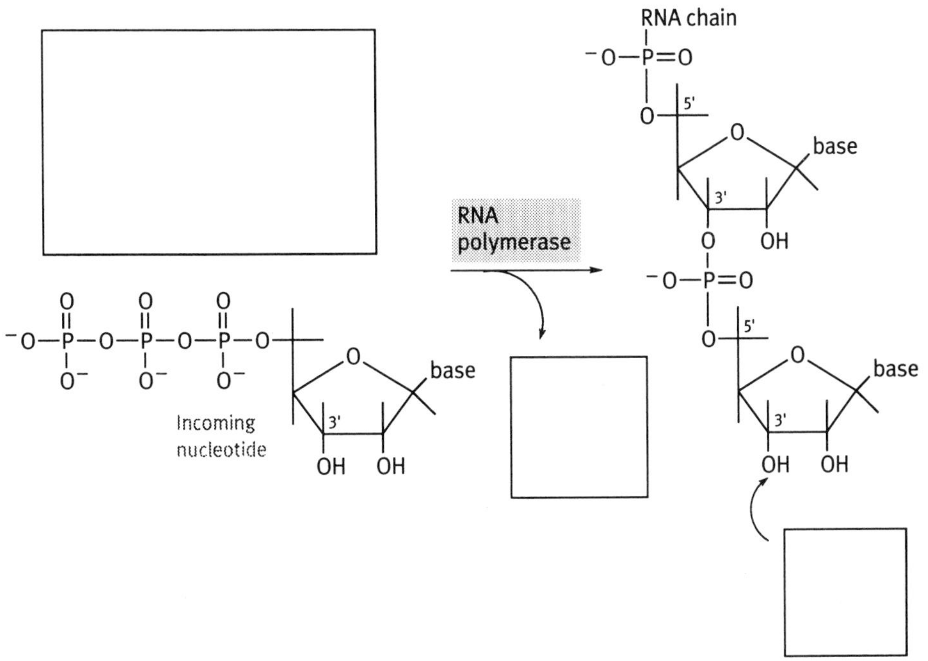

RNA chain

$^-O-P=O$

RNA polymerase

Incoming nucleotide

base

base

base

Transcription terminology

Information flows from DNA to mRNA by a process called ....(1) and from mRNA to protein by a process called ....(2). The sequence of DNA ultimately controls the ....(3) sequence. Only ....(4) of the strands of DNA is transcribed to mRNA. The two DNA strands have complementary base pairs (A–T, G–C) and the mRNA that is made is complementary to the strand that is used as the ....(5). Thus the mRNA that is made and the DNA strand that is ....(6) used as the template (nontemplate) are both ....(7) to the template strand. They have the same sequence (except that U is used instead of T). Thus, to avoid confusion, the nontemplate strand is called the ....(8) or sense strand. A gene is a specific section of DNA that is transcribed into RNA, although not all of this is mRNA. mRNA molecules have sections at each end that are not translated into protein, called ....(9) regions, UTRs. The UTR on the 5′ end of the mRNA contains signals necessary for ....(10) of translation while the 3′ UTR contains signals for ....(11). These are all considered part of the gene. There are other regions of DNA associated with the gene such as the promoter and the terminator. The DNA bases are numbered so that +1 is the first base read, also called ....(12) stream, and ......(13) is the one before this on the ....(14) side of the coding strand, also called upstream. Note that the roles played by the two strands of DNA are not the same for every gene. The ....(15) strand for a certain gene may not be the same DNA strand as the template strand for another gene. Polymerases can read from 'left to right' or 'right to left', but they will necessarily use a ....(16) strand in the DNA duplex. The three phases of gene transcription are ....(17), ...(18), and ....(19). Short stretches of bases are often called ....(20) or elements. A typical *E. coli* promoter has the ....(21) box at −10 and another box at −35. By comparing a number of different Pribnow boxes the ....(22) sequence has been determined to be TATAAT given in the 5′ to 3′ direction of the coding strand. Recognition of these boxes by the RNA polymerase involves the ....(23) factor protein which is only needed during the transcription of the first few bases. The DNA strand is temporarily unwound at the site of a 'bubble' as the polymerase moves along synthesizing the RNA at about 40 nucleotides per second.

Answers: (1) transcription; (2) translation; (3) amino acid; (4) one; (5) template; (6) not; (7) complementary; (8) coding; (9) untranslated; (10) initiation; (11) termination; (12) down; (13) −1; (14) 5′; (15) template; (16) different; (17) initiation; (18) elongation; (19) termination; (20) boxes; (21) Pribnow; (22) consensus; (23) sigma.

Complete the following

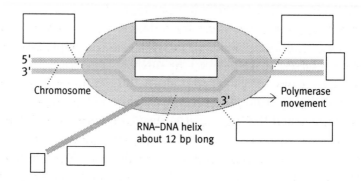

Chromosome

5′
3′

RNA–DNA helix
about 12 bp long

3′

Polymerase
movement

Control of transcription

One method by which ....(1) transcription is specifically ....(2) involves the ....(3) factor which attaches to the newly transcribed mRNA and moves along behind the RNA polymerase. This protein unwinds the RNA–DNA ....(4) by a ....(5) activity as it moves along the chain, and termination of synthesis occurs when this Rho factor catches up with the ....(6). Genes that are 'switched on' all the time are referred to as being ....(7) expressed. Promoters fall into the categories of being weak or strong depending on the ....(8) with which transcription is initiated, and this reflects the ....(9) of mRNA molecules that are made. The base sequences of the Pribnow and −35 boxes as well as the ....(10) between them and the nature of the ....(11) between them influence the strength of a given ....(12). There are different sigma factors, which cause different genes to be active in response to certain environmental conditions such as heat shock, nitrogen ....(13), and other adverse conditions. The production of a protein in response to a chemical signal is called ....(14), and the chemical is called the ....(15), while the prevention of the production of a protein is called ....(16). Many prokaryotic genes are grouped together, with the individual groups being under the transcriptional control of a single promoter. mRNA from this type of gene is called ....(17), and the group of genes including the promoter is called an ....(18).

Answers: (1) prokaryotic; (2) terminated; (3) Rho; (4) duplex; (5) helicase; (6) polymerase; (7) constitutively; (8) frequency; (9) number; (10) distance; (11) bases; (12) promoter; (13) starvation; (14) induction; (15) inducer; (16) repression; (17) polycistronic; (18) operon.

The *lac* and *trp* operons

There are three special proteins that an *E. coli* cell would need to induce in order to survive if lactose were the sole source of carbon. They are ....(1), β-galactoside ....(2), and galactoside ....(3). The DNA codes for these three proteins along with a protein called the *lac* ....(4) protein which can bind to a portion of the DNA just ....(5) stream of the promoter in this operon called the ....(6). In addition, there is a section of DNA upstream of the promoter called the ....(7) binding site, catabolite gene ....(8) protein. The operator and the CAP binding site are places on the DNA where specific proteins can interact. The CAP protein binds to this CAP binding site in response to high levels of ....(9) which reflect ....(10) levels of glucose in the cell. The binding of the CAP protein changes the *lac* ....(11) from a weak one into a high-strength promoter. However, a low level of glucose is not in itself a sufficient signal to turn on this ....(12) because, unless lactose is also present, the *lac* ....(13) protein binds to the operator and the polymerase cannot transcribe the rest of the genes. A ....(14) of lactose interacts directly with the *lac* repressor protein causing it ....(15) to interact with the operator. This metabolite of lactose is the sugar ....(16); it is called the inducer.

When this metabolite is no longer present, the *lac* repressor binds to the operator and ....(17) transcription.

There is also a tryptophan operon (*trp* operon), which is similar to the *lac* operon, but also includes a mechanism known as ....(18) which works to stop the transcription. The mechanism is based on the fact that, in prokaryotes, the ....(19) of mRNA into protein begins as soon as even a small stretch of the mRNA is made. In this case a 14-amino-acid peptide, called the ....(20) peptide, is made first. The ability of the cell to make this leader peptide is used as a ....(21) to attenuate the synthesis of proteins that ....(22) tryptophan. If there is very little tryptophan available, then, as the operon begins making mRNA and the mRNA is being made into the leader protein, the ribosome that is involved in making the protein will ....(23) at a certain place on the mRNA where ....(24) tryptophan amino acids need to be added. This hesitation ....(25) the partially completed mRNA molecule from forming a ....(26) struc-ture, which would, if formed, ...(27) the process of mRNA synthesis. Con-versely, when there is an abundance of tryptophan, the ribosome does .....(28) hesi-tate, the stem loop structure ....(29) form, and the rest of the ....(30) is *not* made.

Answers: (1) β-galactosidase; (2) permease; (3) transacetylase; (4) repressor; (5) down; (6) operator; (7) CAP; (8) activator; (9) cAMP; (10) low; (11) promoter; (12) operon; (13) repressor; (14) metabolite; (15) not; (16) allolactose; (17) stops; (18) attenuation; (19) translation; (20) leader; (21) signal; (22) synthesize; (23) hesitate; (24) two; (25) prevents; (26) stem loop; (27) stop; (28) not; (29) does; (30) mRNA.

Complete the following

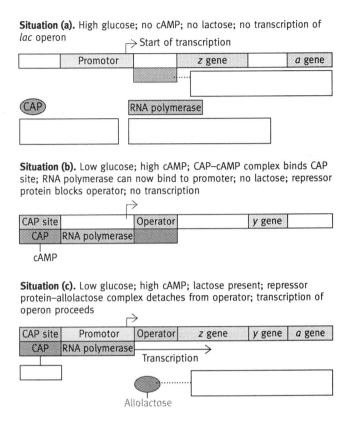

Situation (a). High glucose; no cAMP; no lactose; no transcription of *lac* operon

Situation (b). Low glucose; high cAMP; CAP–cAMP complex binds CAP site; RNA polymerase can now bind to promoter; no lactose; repressor protein blocks operator; no transcription

Situation (c). Low glucose; high cAMP; lactose present; repressor protein–allolactose complex detaches from operator; transcription of operon proceeds

The eukaryote system

The eukaryotic system shares many principles of operation with the ....(1) one. The ....(2) step is very much more complex and inseparable from the subject of gene control. The details of the ....(3) signal are not well understood at all, except that the enzyme responsible for making the mRNA, RNA ....(4), synthesizes a specific AAUAAA sequence after the end of the gene and then goes on to stop at a very distant spot. The mRNA is further processed and a ....(5) tail is added. As it is made, the RNA is ....(6) at its 5′ end, the first base in the strand. This base still has a ....(7) group on the 5′ hydroxyl with the 3′-OH is linked to the rest of the chain. The capping involves the addition of a ....(8) residue making a ....(9) triphosphate linkage, and the G base of the cap is then ....(10) in the N-7 position as is the 2′-OH of the second and sometimes the third nucleotide. The cap is believed to ....(11) the end from ....(12) attack and is involved in the initiation of translation.

The RNA coding for proteins in the eukaryotic system is split up into several parts linked together by stretches of RNA that do ....(13) code for proteins. The sections of the DNA that make these stretches of RNA that do not code for proteins are called ....(14), while the coding sections are called ....(15). Introns can be numerous (2–50) and lengthy (20–20 000 base pairs) and are transcribed in the ....(16) transcript. By a process known as mRNA ....(17), the ....(18) are cut out and the exons are ....(19) together by a mechanism involving a transesterification reaction in which no significant chemical energy is lost. All introns begin with GU and end with AG, and are thus identifiable to the enzymes of the splicing system. The mechanism begins with a 2′-OH of an adenine in the intron chain which attacks the 5′ phosphate of the G nucleotide at the splice site forming a ....(20). The now free 3′-OH end of the first exon attacks the 5′ phosphate end of the second exon and the two exons are joined. This sequence of events is catalysed by ....(21), which contain both proteins and RNA, the latter being referred to as ....(22), small nuclear RNA. Interestingly, it was recently discovered that the primitive organism *Tetrahymena* can splice out its ....(23) without an enzyme, the first discovery of a reaction catalysed by an ......(24) molecule rather than by a ....(25). Introns often code for discrete protein domains, and it is thought that previously unknown proteins can be made by the organism by shuffling these ....(26), where domains would confer a partial function. The mechanism would involve ....(27) shuffling and gene ....(28). Two models to explain the origin of split genes are the 'introns late' and 'introns early' models. The examination of genes that appear to have existed unchanged over long periods of time indicated that there was .....(29) correspondence between the intron–exon structure and the structural domain features of the protein and thus does not support the ....(30) theory. mRNA splicing can also occur in different patterns (rather than linear) and thus produce alternative splicing patterns that lead to the production of different proteins of the same gene.

Answers: (1) prokaryotic; (2) initiation; (3) termination; (4) polymerase II; (5) polyA; (6) capped; (7) triphosphate; (8) GMP; (9) 5′–5′; (10) methylated; (11) protect; (12) exonuclease; (13) not; (14) introns; (15) exons; (16) primary; (17) splicing; (18) introns; (19) spliced; (20) lariat loop; (21) spliceosomes; (22) snRNA; (23) introns; (24) RNA; (25) protein; (26) domains; (27) exon; (28) crossovers; (29) no; (30) introns early.

Complete the following

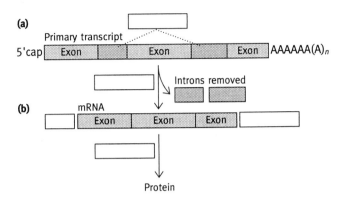

Eukaryotic control mechanisms

Gene transcription and control in the eukaryotic system begins with the
.......................(1) of the DNA by selectively loosening up the pieces of DNA to
be transcribed. Evidence for this comes from the facts that chromosome ....(2)
have been observed in the salivary gland cells of the larval form of the fruit fly and
that ....(3) active regions of chromatin become more susceptible
to attack by DNA nucleases, DNase. There are about 50 000–100 000 ....(4)
in a human, but only 4000–5000 in *E. coli*, with most of the *E. coli* expression
occurring during the time of division. Animal systems coordinate cell activities that
call for a modulation of the synthesis of specific proteins via the control of gene
.......................(5). The complexity of the controls is very great and thus a given
gene may need to be ....(6) to a number of different signals. In
addition, there is a need to express different genes in different ....(7).
Genes expressed in all tissues at all times are called ....(8) genes.
And, at another level, the expression of certain genes is ....(9)
specific (cf. embryonic versus mature). Eukaryotic genes have a promoter upstream
from the initiation region and, in about 80% of the cases, a ....(10) box at
about −25. Further upstream, within about 100–200 base pairs, are regions of the
DNA (boxes) that are recognized specifically by proteins called ....(11)
factors as well as regions called ....(12) that bind protein ....(13),
which can often be located quite far (1000 base pairs) away from the initiation region. These enhancers are ....(14) strictly included as part of the ....(15)
since it is not in a relatively fixed position with respect to the start site. There is a
great deal of variability in the way eukaryotic genes are controlled since no one element is essential for transcription and different genes have different combinations of
elements. Eukaryotic gene transcription is initiated by a group of proteins known as

the general transcription factors which form the ....(16) initiation complex on the DNA to which the RNA polymerase attaches and which is needed for the transcription of all type II genes. The TATA binding ....(17), TBP, and the TBP-associated factors, ...(18), form a complex known as the transcriptional factor D for polymerase II, ...(19), which is joined by RNA polymerase II and other proteins to complete the basal ....(20) complex. Genes without a TATA box and those transcribed by polymerase I and III are initiated by other related mechanisms. The specific transcription factors ....(21) combine with the upstream and enhancer elements as long as the DNA is associated with ....(22) in the structures of the ....(23). The nucleosome must be displaced before the basal initiation complex can assemble. Transcriptional factors that bind to elements distant from the initiation region can participate in a complex by a ....(24) action of the DNA. There are at least two binding sites on each ....(25)—one to interact with the DNA and the other to interact with the other proteins of the transcriptional complex. This mechanism for eukaryotic gene control is enormously ....(26), since many factors can influence gene expression. Some major questions about how the interactions between transcriptional factors and the initiating complex control initiation are yet to be answered.

Answers: (1) unpacking; (2) puffs; (3) transcriptionally; (4) genes; (5) transcription; (6) responsive; (7) tissues; (8) housekeeping; (9) developmentally; (10) TATA; (11) transcriptional; (12) elements; (13) enhancers; (14) not; (15) promoter; (16) basal; (17) protein; (18) TAFs; (19) TFIID; (20) initiation; (21) cannot; (22) nucleosomes; (23) chromatin; (24) looping; (25) protein factor; (26) flexible.

Complete the following

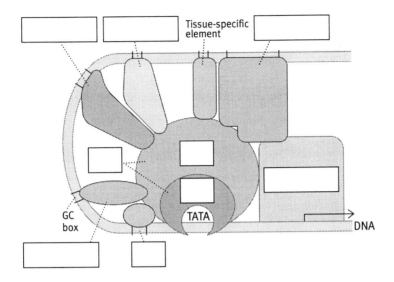

DNA binding proteins

Most ....(1) that bind to specific segments of ....(2) can be grouped into a small number of families based on their structural characteristics. They are the helix–turn–helix, the ....(3), the leucine ....(4), and the zinc ....(5) proteins. These proteins bind to double-stranded DNA, primarily to the bases in the major groove of the double helix. There are often two adjacent sites on the DNA necessary for complete attachment, and the protein is usually a ....(6). It is important that the binding of these proteins to their specific DNA region be ....(7), and thus occur (or not occur) only in response to some ....(8), usually from a hormone. In the helix–turn–helix, ...(9) motif, the recognition helix sits in the major groove and the DNA recognition site, which is also the operator, has an imperfect ....(10) symmetry that constitutes two identical binding sites. Each of the monomers of the DNA binding protein has an HTH so that interaction with one of the binding sites is made up by ....(11) of the palindromic sequence; the other half of the palindrome makes up the other receptor, which interacts with the other subunit in the dimer. Note that only one of the helices in each of the HTHs interacts with the DNA, and that the dimer has ......(12) HTHs. The palindromic symmetry of the DNA shown on page 286 of the main text can be seen in the following way. Starting at the bases marked with the blue dotted line, read to the right on the top strand ....(13) and to the left on the bottom strand CGGTGATA; then notice that the bases typed in black in these two lists are the same. Some are in red to indicate that the palindromic symmetry in this example is not perfect. Since the binding proteins recognize the specific DNA sequence in either direction there are two sites, one on each strand. The ....(14) zipper proteins contain the dimeric structure in which the whole molecule is an α helix in which every seventh amino acid is a leucine. Since there are ....(15) residues per turn, all the leucines appear on the same side of the α helix forming a ....(16) face; it is through this face that the two monomers interact to form the dimer. The dimers attach to the DNA in a ....(17) grip, with the two arms being in adjacent major groove sections of the duplex. The leucine residues do not interdigitate as the name would imply. In the zinc finger proteins a zinc atom attaches to two ....(18) and two ....(19) to form a finger-like structure, one part of which is a recognition α helix that sits in the major groove of DNA. Some Zn fingers have more than one such finger, even up to 30.

Answers: (1) proteins; (2) DNA; (3) homeodomain; (4) zipper; (5) finger; (6) dimer; (7) controllable; (8) signal; (9) HTH; (10) palindromic; (11) half; (12) two; (13) CAGTGGTA; (14) leucine; (15) 3.6; (16) hydrophobic; (17) scissors; (18) histidines; (19) cysteines.

Homeodomains

The homeodomain proteins are important in one of the biggest remaining unsolved areas of biology, the mechanism of differentiation or embryological development. As cells differentiate into various tissues, selected genes are expressed in different tissues and relatively little is known about how this happens. Most of the experiments have been done on the fruit fly *Drosophila*. Homeodomains are proteins of about 60 amino acids in length that recognize certain DNA domains and function as transcription factors. Homeodomains are highly conserved, being similar in structure in the insect and in vertebrates, including humans. The major function of homeotic genes in insects is to regulate the expression of other genes, which in turn determine the features characteristic of each body segment. This is the result of unique expression domains or combinations of overlapping expression domains. It has been shown that the protein encoded by the homebox enables a number of gene regulators to recognize and bind to the gene under their control. Proteins made by a homeotic gene are found in specific segments of the insect's body.

mRNA stability and other topics

The stability of individual mRNAs also has an effect on the ....(1) of ....(2) synthesis since the level of mRNA in a cell is a function of its ....(3) and ....(4) rates. A messenger with a long half-life will be present at a higher steady-state level and result in a ....(5) rate of synthesis of the encoded protein. Gene expression can thus be regulated by altering the ....(6) of a given mRNA. The rapid messenger turnover with half-lives on the order of

...................(7) for prokaryotes and between 10 minutes and 2 days for mammals permits rapid ....(8) to changing circumstances. The low stabilities of mRNAs that code for ....(9) proteins give a rapid control response. There is thus a wide range of stabilities of different mRNAs and these can ....(10) within a cell in response to changing conditions. Most eukaryotic mRNAs have a ....(11) tail attached to the 3′ end, which is thought to offer protection against breakdown. Deadenylation is thought to be a factor in this process. ....(12) mRNA offers a number of exceptions to the general rule. It is required only during the S phase of the eukaryote cell cycle, and it appears that free histone monomers are involved in the mechanism for the switch-off of histone synthesis, which is necessary at the end of S phase, with the lifetime of the mRNA playing an important role. The synthesis of the protein β-tubulin is also regulated by a connection between the ....(13) of the monomer form of this protein and the ....(14) of its mRNA.

Transferrin, a blood plasma protein that transports iron, has a receptor protein. Its synthesis is regulated by the stability of its ....(15). In the 3′ untranslated region of the mRNA there is a group of five ....(16), called iron-responsive elements, ...(17). There is an IRE binding protein that attaches and stabilizes the mRNA, but only in the ....(18) of iron. It appears that some mRNAs are made less stable by the presence of an adenine/uracil-rich element (AURE), which interacts with a specific binding protein. There is evidence that the loss of this ....(19) may be involved with genes becoming ....(20) producing.

In mitochondria (as with chloroplasts), which have their own ....(21) and protein-synthesizing machinery, there are ....(22) copies of circular DNA. However, the mitochondria do ......(23) have all of the DNA needed to completely reproduce, and the nucleus of the cell is involved with the rest. Interestingly, the mechanism of protein synthesis has many prokaryotic characteristics, supporting the theory that mitochondria evolved from ....(24) prokaryotic cells that became ....(25). There is also evidence for direct ....(26) of mRNAs by inserting or changing bases within the coding region. This is an unprecedented concept since the changes in the mRNA do not appear to be coded for in the ....(27). There are also genes that code for special RNA molecules that are not messengers, notably the ....(28) and rRNA. These genes have special polymerase enzymes for their synthesis, ...(29) for most of the rRNA and Pol III for tRNA and the smaller rRNA. In some of these, the regulatory elements can occur ....(30) of the start site.

Answers: (1) rate; (2) protein; (3) synthesis; (4) breakdown; (5) higher; (6) half-life; (7) 2–3 minutes; (8) responses; (9) regulatory; (10) change; (11) polyA; (12) Histone; (13) amount; (14) stability; (15) mRNA; (16) stem loops; (17) IRE; (18) absence; (19) AURE; (20) cancer; (21) DNA; (22) multiple; (23) not; (24) engulfed; (25) symbiotic; (26) editing; (27) DNA; (28) tRNAs; (29) Pol I; (30) downstream.

Review of problems from the end of Chapter 21

- These are the three main differences between RNA synthesis and DNA synthesis: RNA contains the base U while DNA contains T; RNA polymerase does not require a primer as DNA polymerase does; and RNA synthesis does not include proofreading.

- The DNA contains a promoter region upstream and a termination site downstream of the start site of transcription. The sequence of the mRNA that is made can be deduced directly from the sense strand of DNA which is complementary to the template actually used to make the mRNA. There are UTR regions on the mRNA on both the 5′ and 3′ ends, which are involved with the initiation and termination of translation, respectively.

- The DNA polymerase along with a sigma factor interact with the DNA at the Pribnow and the −35 boxes.

- One method by which *E. coli* gene transcription is terminated involves a G–C stem loop and the other involves the Rho factor.

- There appears to be a number of factors related to the specific base sequence just upstream and downstream of the initiation site that determine the strength of a promoter.

- The *lac* operon controls the expression of genes that code for enzyme proteins that are needed for the metabolism of lactose. The expression depends on the levels of glucose, cAMP, and lactose and on the inducer molecule, allolactose. The process involves the catabolite gene activator protein (CAP) and its DNA binding site, the operator DNA segment, a *lac* repressor protein that binds to the operator in the absence of the inducer allolactose, the promoter portion of the DNA, and an RNA polymerase enzyme. For transcription to proceed, cAMP, the CAP, allolactose, and the polymerase must be present.

- The eukaryotic system is more complicated than that in prokaryotes with the inclusion of introns and 5′ capped and 3′ polyadenylated ends.

- Splicing involves cutting out the introns. This process is related to exon shuffling, which may be involved in the process in which a single gene gives rise to different proteins.

- The initiation of transcription in eukaryotes is more complex than in prokaryotes, involving a much greater number of transcriptional factors.

- The four families of transcriptional factors based on structural motifs presented in this chapter are helix–turn–helix, leucine zipper, zinc finger, and homeodomain proteins.

Additional questions for Chapter 21

1. When a new base nucleotide is added to the growing RNA chain, what is the other product of this reaction and why is it important?

2. In viral transcription, which of the DNA strands is referred to as the minus (–) strand?

3. What is the function of the G–C rich stem loop structure in the termination of transcription?

4. What is the nature of the reaction catalysed by the enzyme β-galactosidase?

5. What does CAP stand for and what role does it play in the *lac* operon?

6. What is the chemical nature of the 5′ cap in eukaryotic mRNA and what effect does this structure have?

7. What is the name of the protein–RNA species that catalyses the splicing reaction?

8. What are TBP, TAFs, and TFIID and how are they involved in the transcription of type II eukaryotic genes?

9. List three types of control elements that can influence the initiation of eukaryotic gene transcription.

10. Which of the following represent a palindromic sequence:
    (a) —GGGGCCCC—;
    (b) —ATCGTAGC—;
    (c) —ATCGCGAT—?

11. What constitutes the 'teeth' in the leucine zipper motif?

12. How is mRNA stability related to gene expression?

13. List four structural motifs found in the 3′ untranslated regions of mammalian mRNAs that influence the half-lives of those molecules in the cell.

14. Which two amino acids play important roles in the sequence of the zinc finger proteins?

15. Name two types of RNA molecules that do not code for proteins.

Chapter 22

....

Protein synthesis, intracellular transport, and degradation

Chapter summary

This chapter describes the processes by which proteins are made, transported, and degraded within cells. The prokaryotic and eukaryotic systems for translation of the mRNA into protein and the roles of initiation factors, the Shine–Dalgarno sequence, and elongation factors are presented and compared. The factors that influence protein folding and the occurrence of prions, which are associated with abnormal conformations, are important in the process of making proteins. The roles of the endoplasmic reticulum and the Golgi apparatus in processing and sorting proteins are presented. The chapter ends with a description of protein turnover and the important role of ubiquitin in this process.

Learning objectives

- ❏ The importance of the three-base codon.

- ❏ The fact that most amino acids have more than one codon directing their synthesis.

- ❏ The role of the transfer RNA.

- ❏ The implications of wobble pairing.

- ❏ The occurrence of a special initiation codon and initiation factor IF2 in the prokaryote system.

- ❏ The transformylase activity that uses the N^{10}-formyltetrahydrofolate cofactor to generate formylated methionine, fMet.

- ❏ The importance of the Shine–Dalgarno sequence of bases.

- ❏ The relationships among the 50S, 30S, and 70S ribosomal forms and the importance of GTP hydrolysis and the various initiation factors.

- ❏ The importance of the GTPase activity of the G proteins.

- ❏ The roles of the elongation factors, EF-Tu and EF-G and the G proteins in moving the ribosome along the mRNA in the process of translocation.

- The direction of protein synthesis.

- The events that occur at the termination of the message on the mRNA.

- The different means used by the prokaryote and eukaryote systems to ensure that the mRNA is positioned correctly with respect to the AUG initiation codon.

- The effect of antibiotics on the various components of the translation process.

- The roles of protein disulfide isomerase (PDI), peptidyl proline isomerase (PPI), and the molecular chaperones in protein folding.

- The role of prions in certain neurological degenerative diseases and their relationship to abnormal protein folding.

- The role of the endoplasmic reticulum (ER) for newly synthesized proteins that are destined for other than the cytoplasm, nucleus, or mitochondria.

- The activity of the peptidyl transferase structure.

- The roles of the protein leader sequence, the signal recognition particle (SRP), the translocon, and signal peptidase in transferring the appropriate proteins through the ER membrane.

- The mechanism responsible for the glycosylation of proteins.

- The role of the Golgi complex in sorting and sending proteins to their appropriate destination.

- The mechanisms for handling integral membrane proteins and the roles of the leader and anchor sequence signals.

- The distinction between pre- and pro-proteins.

- The importance of protein turnover with regard to removing damaged proteins and as an aspect of the metabolic control mechanism and the role that ubiquitin plays in this process.

A walk through the chapter

Protein synthesis

The linear ....(1) of the four different bases in mRNA are used to construct ....(2) from the 20 different ....(3) by using a ....(4) base code for each amino acid. The triplet of bases is called a ....(5). The genetic code directs the ....(6) of the base pairs into amino acids. Not all three-base combinations of the four available bases correspond to an amino acid. Three of these codons are stop signals. There is a ....(7) in the system such that most, all except ....(8) and ....(9), of the amino acids have more than one codon directing their inclusion in the ....(10). Codons for amino acids that have similar ....(11)   tend to have similar three-letter codes. The appropriate amino acid is recognized by the codon on the mRNA by a ....(12), tRNA, specific for that amino acid. tRNA molecules contain a three-base ....(13),   which base pairs (A with U, and G with C) to the codon on the ....(14). Each tRNA carries the amino acid that corresponds to the codon with which the tRNA will base pair in accord with the translated ....(15).   Some tRNA molecules can recognize ....(16)   than one of the codons that specify for that amino acid by a mechanism known as ....(17)   pairing. The correct pairing of the ....(18)   two correctly is enough to override an incorrect pairing of the ....(19). The sequence in an anticodon is given in the 5′ to 3′ direction. The codon–anticodon interaction is antiparallel so the base that can ....(20) in the codon sequence is the last of the three written in the ....(21) direction. This is the way in which tRNAs are usually displayed. The benefit from this wobble pairing is that it permits the cell to synthesize ....(22)   species of tRNA molecules. Another variation on this theme is the use of the base ....(23) in the anticodon, which can pair with C, U, or A nucleotide bases. Note that hypoxanthine ribose is also called ....(24).

Answers: (1) sequences; (2) proteins; (3) amino acids; (4) three-; (5) codon; (6) translation; (7) degeneracy; (8) methionine; (9) tryptophan; (10) protein; (11) structures; (12) transfer RNA; (13) anticodon; (14) mRNA; (15) genetic code; (16) more; (17) wobble; (18) second; (19) first; (20) wobble; (21) 5′ to 3′; (22) fewer; (23) hypoxanthine; (24) inosine.

Complete the following

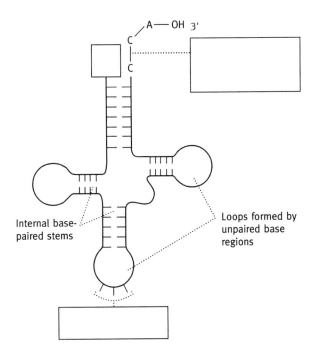

Internal base-paired stems

Loops formed by unpaired base regions

Aminoacyl-tRNA synthetase

tRNAs are only charged with the amino acids that correspond to the proper translation code. The three letters of the ....(1) are included as a superscript to tRNAs (cf. tRNA^{Phe}). Enzymes that attach amino acids to tRNAs are called
aminoacyl-tRNA ....(2) and employ ....(3) and pyrophosphate ....(4) to drive the reaction. Each one of these is ....(5)
for the tRNA that carries a certain amino acid. A ....(6)   mechanism involving the aminoacyl-AMP intermediate formed in the synthetase reaction
exists, such that, if at this stage of the activation an ....(7)   amino acid
has been added, the error is corrected by hydrolysing the bond between this incorrect amino acid and the ....(8).   This process reduces the error rate down to 1
in several ....(9).   When the tRNA is ....(10),   the amino acid is
attached to the free ....(11) group on the terminal adenine nucleotide residue of
the tRNA via an ....(12) bond involving the ....(13) of the amino
acid. This bond has about the same energy content as a ....(14)   bond
so that the transfer of this amino acid to the ....(15) group on the end of
the growing polypeptide chain is facilitated.

Answers: (1) amino acid; (2) synthetases; (3) ATP; (4) hydrolysis; (5) specific; (6) proofreading; (7) incorrect; (8) tRNA; (9) thousand; (10) charged; (11) 3′-OH; (12) ester; (13) carbonyl; (14) peptide; (15) amine.

Ribosomes

Ribosomes are called particles because of their ....(1) and detectability via ....(2). They are mutisubunit particles made predominantly (60%) from ribosomal RNA, ...(3), with the remainder coming from a number of different ....(4). The single strand of rRNA forms numerous internal ....(5) pairings. The size of a particle this big is expressed in ....(6) units (S) which is a measure of how ....(7) the material sediments in an ultracentrifuge. Proteins are synthesized by the action of the ....(8) moving down the mRNA (5′ to 3′), while the ....(9), charged with ....(10) acids, file in according to the code of the tRNA and deliver their amino acids to the growing chain. This continues until the ribosome hits a ....(11) codon and the protein is released and the ribosomes dissociated into subunits. The ....(12) of translation occurs after the 5′ untranslated region of the mRNA and depends on the ribosome being in the correct ....(13) frame, that is, with the correct three-base grouping down the chain. A reading frameshift error would make the combination of three bases either ....(14) or ....(15). This can also happen if a base on the ....(16) is deleted or added. The prokaryotic system contains a ribosome positioning section 5′ of the translational start site. The start codon is ....(17), although sometimes GUG is used. AUG also codes for methionine. There are two different ....(18) for methionine and one is only used for ....(19) purposes. The tRNA for methionine that initiates translation can base pair with either AUG or GUG due to a ....(20) at the 5′ base of the ....(21). The selective use of this tRNA for initiation is due to the action of an initiating protein factor, ...(22). In addition, the methionine that becomes attached to the initiating tRNA is ....(23) on its NH_2 group by an enzyme with ....(24) activity using N^{10}-formyltetrahydrofolate as the ....(25). The formyl group, and frequently the methionine also, are ....(26) before completion of the synthesis of the protein. When a formyl group is attached to the methionine and/or this formylated methionine is attached to its transfer RNA, an 'f' is included as in $fMet\text{-}tRNA_f^{Met}$.

Answers: (1) size; (2) microscopy; (3) rRNA; (4) proteins; (5) base; (6) svedberg; (7) fast; (8) ribosomes; (9) tRNAs; (10) amino; (11) stop; (12) initiation; (13) reading; (14) wrong; (15) nonsense; (16) mRNA; (17) AUG; (18) tRNAs; (19) initiation; (20) wobble; (21) anticodon; (22) IF2; (23) formylated; (24) transformylase; (25) cofactor; (26) removed.

Initiation

The initiation factor IF3 binds to the 30S subunit and helps to keep it from remaining bound to the 50S after the ribosome completes the run through a section of mRNA. IF1 and IF2 are also involved. The Shine–Dalgarno sequence of bases on the mRNA is complementary to a section of the 10S rRNA (r for ribosomal) and this helps to position the mRNA on the small ribosomal subunit. The 50S subunit associates with the mRNA and 30S subunit after the IF3 is released. There is a GTP hydrolysed and the IF1 and IF2 are released, leaving a complete 70S ribosome positioned on the mRNA in the P site with the donor fMet-tRNA$_f$ and a vacant A site awaiting the delivery of the second amino acid on its tRNA. In cases where the mRNA molecule encodes for more than one protein, known as polycistronic mRNA, each coding region has its own Shine–Dalgarno sequence.

Complete the following

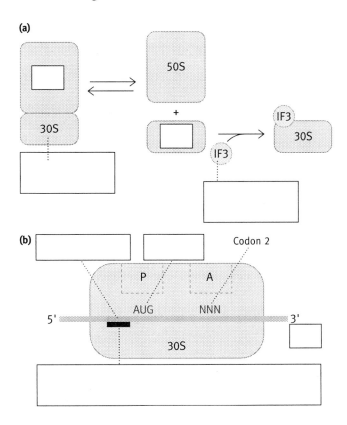

The mechanism of translation

The mechanism of translation involves three sites on the ribosome. E, P, and A referred to as the exit, peptidyl, and acceptor sites, respectively. Synthesis begins with the P site aligned with the ....(1)   codon number one. The incoming ....(2)   attaches to the A site. The new peptide bond is formed with the growing chain attached to the ....(3)   through the most recent tRNA. As translocation occurs, the previous tRNA exits from the .....(4) site, the growing chain finds itself in the .....(5) site, and the ......(6) site is ready for the next aminoacyl-tRNA. Two proteins known as ....(7)   factors are also involved; they are members of a class known as the 'G' proteins, which bind with ....(8)   and have ....(9)   activity. They interact with the ribosome and, when the bound GTP is ....(10),   a ....(11) change in the ribosome occurs. Specifically, ...(12)   (elongation factor, temperature unstable) aids in delivering the incoming aminoacyl-tRNA to the ribosome, while ....(13) moves the ....(14) along the mRNA in the ....(15)   direction. The energy of the GTP hydrolysis drives the ....(16).   The .....(17) site is only used for the initiation process involving the fMet-tRNA$_f$. The subsequent aminoacyl-tRNAs arrive at the ......(18) site together with the ....(19)   and a molecule of ....(20).   The fMet - tRNA$_f^{Met}$ does ......(21)   have an EF-Tu-GTP bound to it. The enzyme peptidyl ....(22)   transfers the fMet group to the free ....(23) group of the aminoacyl-tRNA in the A site. The ribosome now moves along the mRNA to the next set of three bases (codon) by a process called ....(24)   requiring EF-G (also known as translocase) which hydrolyses GTP. Two models have been proposed to account for the details of this process. It is clear that the growing peptide remains in a ....(25)   position relative to the movement of the ribosomes. The process continues with a new ....(26) arriving with EF-Tu-GTP at the A site, with transfer of the growing peptide chain and movement of the ribosome with the help of ....(27).   The synthesis proceeds N-terminal to C-terminal until one of three stop codons is reached (UAG, UAA, or UGA). Then, in a step that involves a specific protein ....(28) factor, the finished protein is released from the tRNA by a simple hydrolysis of the ....(29)   bond between the ....(30) amino acid and the 3'-OH of the tRNA. The ribosome detaches from the mRNA and dissociates into subunits.

Answers: (1) mRNA; (2) tRNA; (3) mRNA; (4) E; (5) P; (6) A; (7) elongation; (8) GTP; (9) GTPase; (10) hydrolysed; (11) conformational; (12) EF-Tu; (13) EF-G; (14) ribosome; (15) 5' to 3'; (16) elongation; (17) P; (18) A; (19) EF-Tu; (20) GTP; (21) not; (22) transferase; (23) amino; (24) translocation; (25) fixed; (26) aminoacyl-tRNA; (27) EF-G-GTP; (28) release; (29) ester; (30) C-terminal.

Complete the following

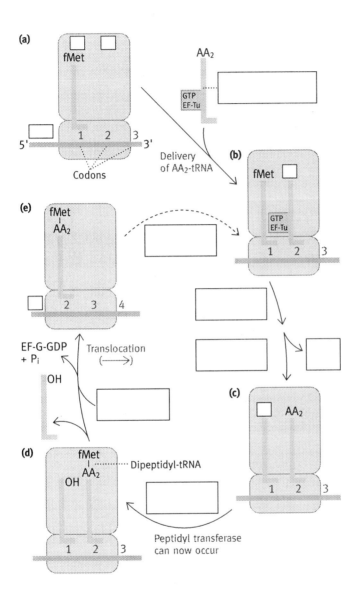

Specifics of the eukaryote system

When more than one ribosome synthesizes protein from a given mRNA, this structure is called a.....................(1). ....(2)   have larger ribosomes
and, although methionine is always the first amino acid, it is not ....(3),
although there is a special methionyl-tRNA used only for initiation. Although a
number of details between the prokaryotic system and the eukaryotic system are the

same, the mechanism by which the mRNA is ....(4) correctly so that the P site corresponds with the ....(5) initiation codon is completely different. Since the eukaryotic mRNAs are ....(6) with a methylated guanine nucleotide at the ......(7) end and since there is ......(8) Shine–Dalgarno sequence, a group of protein factors is attached to the cap joined with a 40S ribosomal subunit called a ....(9)   complex which then moves along the mRNA until it encounters the first AUG triplet. Then the ....(10) subunit joins in with ....(11) hydrolysis occurring. Since there can be only ....(12) initiation codon per mRNA, they must be monocistronic. Elongation factors EF1α and EF2 are also involved. Antibiotics are a class of molecules that interact with highly important points in the cellular processes. The following table shows some of the points of action of some common antibiotics and toxins.

Antibiotic/toxin	Step affected
Streptomycin	.............................(13)
Kirromycin	EF-Tu release
......................(14)	Peptidyl transferase
......................(15)	Peptidyl transferase
Fusidic acid	....(16)
Diphtheria toxin	Eukaryotic translocase
......................(17)	Inactivator of a ribosome subunit

Answers: (1) polysome; (2) Eukaryotes; (3) formylated; (4) positioned; (5) AUG; (6) capped; (7) 5′; (8) no; (9) pre-initiation; (10) 60S; (11) GTP; (12) one; (13) Initiation; (14) Erythromycin; (15) Chloramphenicol; (16) Translocation; (17) Ricin.

Complete the following

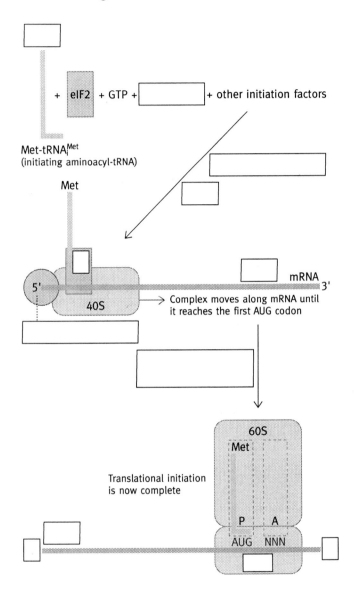

Protein folding and prions

The ....(1)   acid sequence of a protein determines its final native ....(2),   as   was demonstrated with the experiment using the enzyme ribonuclease. Since the protein has to fold in just the right way in a matter of ....(3), it is believed that certain sections of a polypeptide may rapidly assume a secondary structure and that these somehow ....(4)   correct folding of the entire molecule. There is evidence for the involvement of two classes of proteins. The first includes protein ....(5) isomerase (PDI) which acts to break and reform ....(6) bonds until the correct linkages are formed. The other member of this class is peptidyl ....(7) isomerase (PPI) which catalyses the rearrangement of the configuration around proline from ....(8) and back until the correct one is found. This activity has also been found on the cyclophilins. The second class of proteins is known as the ....(9); these proteins have the ability to recognize and bind to partially ....(10) proteins. They bind to the protein as it emerges from the ribosome and ....(11) improper folding interactions from occurring; they then dissociate from the protein as the correct folding takes place, a step that requires ....(12) hydrolysis. The chaperones affect the ....(13) and ....(14) of the folding. The nature of the final folded form is determined by the amino acid sequence. Prions, a term derived from proteinaceous ....(15)   particles, are responsible for a number of neurological ....(16) diseases and are derived from a ....(17) resistant form of a normal protein (PrP). In mice the normal protein, PrP^c, and the ....(18)   for scrapie, PrP^{sc}, have the same polypeptide ....(19) but have different ....(20) conformations. The PrP^{sc} readily forms aggregates, which result in ....(21) plaques. There is evidence that the PrP^{sc} form causes the normal ....(22) to adapt the abnormal conformation. This abnormally folded form then acts in an ....(23)   manner.

Answers: (1) amino; (2) configuration; (3) minutes; (4) facilitate; (5) disulfide; (6) S—S; (7) proline; (8) *cis* to *trans*; (9) chaperones; (10) unfolded; (11) prevent; (12) ATP; (13) speed; (14) pathway; (15) infectious; (16) degenerative; (17) protease; (18) prion; (19) sequences; (20) folded; (21) amyloid; (22) protein; (23) infectious.

Endoplasmic reticulum

Not all proteins that are synthesized in the cytoplasm stay there. Many must be translocated to cell membranes or vesicles or released out of cell. The endoplasmic reticulum, ...(1), is a complex network of interconnecting structures that appear as a flattened bag in the ....(2) microscope. This membranous structure separates the ER ....(3) from the ....(4). The rough appearance is due to the presence of ....(5), while the lack of these gives rise to the smooth ER. Proteins destined for ....(6) than the cytoplasm, nucleus, or mitochondria are first transported into the rough ER lumen. Proteins destined to follow this path have a ....(7)   sequence of about 25 amino acids on the *N*-terminal that fits a characteristic pattern. The leader sequence is synthesized by a free ribosome, which is immediately recognized by a signal ....(8) particle (SRP) that stops the elongation of the peptide. The leader peptide and the polypeptide as it is forming are transferred ....(9)   the ER membrane.

SRP receptors known as ....(10) proteins use the energy provided by a ....(11) hydrolysis. The polypeptide translocating pore is also called a ....(12). Associated with the pore is a signal ....(13) that hydrolyses off the leader sequence. Glycosylation involves the attachment of ....(14) to a protein which generally occurs through the NH_2 of ....(15) or the —OH of serine and ....(16). *N*-glycosylation occurs in the ......(17), and involves the translocation of the sugars in the form of a ....(18) phosphate. Once inside the lumen of the ER, the protein is ....(19) off into transport ....(20) to be delivered to the Golgi complex by membrane ....(21).

Answers: (1) ER; (2) electron; (3) lumen; (4) cytoplasm; (5) ribosomes; (6) other; (7) leader; (8) recognition; (9) through; (10) docking; (11) GTP; (12) translocon; (13) peptidase; (14) sugar; (15) asparagine; (16) threonine; (17) ER; (18) dolichol; (19) budded; (20) vesicles; (21) fusion.

The Golgi

The Golgi receives proteins from the ....(1) and sends them out by budding off ....(2). The Golgi ....(3) out the proteins and sends them out in vesicles that deliver their contents to the correct place. Proteins for secretion can be released ....(4) as they are produced, for example, ...(5) proteins, or, as in the case of ....(6) enzymes, released in response to some ....(7). The destination signals for the proteins may involve the ....(8) portion of the molecule as is the case with ....(9) proteins or ....(10) that recognize structural or sequenced information about the protein. Lysosomal proteins become ....(11) on the oligosaccharide portion and, after the initial budding from the Golgi, these are fused with other 'sorting vesicles' in which the receptors for the phosphorylated groups are dissociated and ....(12) back to the Golgi, while the proteins are delivered to the lysosomes. Diseases associated with ....(13) in this ....(14)   have serious clinical effects. Proteins destined for the ....(15)   membrane are processed through the ER and Golgi as are.............(16) for this membrane. ....(17) membrane proteins that are destined to stay imbedded in the ....(18) become part of the ......(19) during the pathway to the plasma membrane. These proteins carry an ....(20) sequence in the polypeptide chain that fixes itself in the ....(21) interior of the bilayer. This entire piece of the ER is subsequently ....(22) to the plasma membrane with the same ....(23) as in the ER. Peptides that ....(24)   the membranes have ....(25) internal leaders and anchor signals. Another type of protein translocation involves proteins destined for the ....(26) or chloroplast in plant cells. These are synthesized on free ribosomes as ....(27). Chaperones attach to these pre-proteins and keep them in an ....(28) form. The protein in this extended form can cross the mitochondrial membrane. The process of translocation across the mitochondrial membrane involves membrane protein ....(29). This pathway can also distinguish between proteins that are destined for various parts of the mitochondria.

Answers: (1) ER; (2) vesicles; (3) sorts; (4) continually; (5) serum; (6) digestive; (7) signal; (8) carbohydrate; (9) lysosomal; (10) receptors; (11) phosphorylated; (12) recycled; (13) defects; (14) pathway; (15) plasma; (16) lipids; (17) Integral; (18) bilayer; (19) ER; (20) anchor; (21) hydrophobic; (22) transferred; (23) orientation; (24) crisscross; (25) multiple; (26) mitochondria; (27) preproteins; (28) unfolded; (29) receptors.

Complete the following

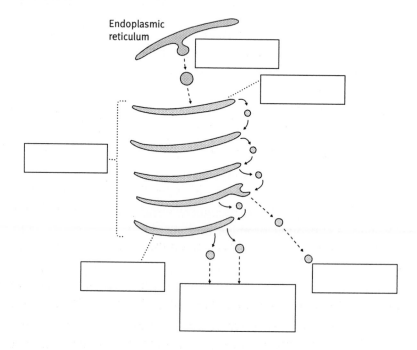

Degradation of proteins

The degradation of proteins occurs in all cells giving rise to protein ....(1) as old protein molecules are replaced with new ones in a highly ....(2) process. The average amount of time between ....(3) and ....(4) can vary greatly. It is important to destroy those proteins that may have been made with the ....(5) protein sequence or that have undergone chemical ....(6). In addition, some proteins are designed to exist for short ....(7) amounts of time. At least a part of the explanation for this turnover of protein involves ....(8)   control. It is possible to sensitively and continuously control the ....(9) of a protein that is being destroyed soon after it is made by adjusting the ....(10)   of synthesis. ....(11)   that play a role in rate-controlling metabolic steps fall into this category, while ....(12) proteins do not. The selective destruction of protein via a mechanism involving the eukaryotic protein ....(13)   is important. When ubiquitin becomes covalently attached to a protein, that protein is ....(14) for destruction.

Answers: (1) turnover; (2) selective; (3) synthesis; (4) degradation; (5) wrong; (6) damage; (7) predetermined; (8) metabolic; (9) level; (10) rate; (11) Enzymes; (12) structural; (13) ubiquitin; (14) marked.

Review of problems from the end of Chapter 22

- The fact that these are 64 codons comes from the possible ways of combining four different things (the bases) in groups of three (the codons). If there were only one codon for each amino acid there would be a great number of combinations not used.

- There is some flexibility in the requirement for exact three-base complementary pairing between the codon and the tRNA. The first base of the anticodon (the 5′ side on the tRNA) does not have to pair exactly. This is known as the wobble mechanism.

- Since the convention for numbering nucleic acids (both RNA and DNA) is 5′ to 3′, these are typically drawn on the page from left to right. For the mRNA to have this directionality, the tRNA must have the opposite.

- Fidelity, the sense of accuracy of reproduction, is based on the correct base pair formation.

- GTP as an energy carrier is involved at three steps: the assemblage of the initiation complex; the delivery of the aminoacyl-tRNAs; and the translocation step.

- E, P, and A refer to the exit, peptidyl, and acceptor sites, respectively. Synthesis begins with the P site aligned with the mRNA codon number one. The incoming tRNA attaches to the A site. The new peptide bond is formed with the growing chain attached to the mRNA through the most recent tRNA. As translocation occurs, the previous tRNA exits from the E site, the growing chain finds itself in the P site, and the A site is ready for the next aminoacyl-tRNA. The straddle model takes into account that the peptide can always remain in the same position relative to the large ribosomal subunit.

- Prokaryotic initiation permits more than one start site so that each individual mRNA of the polycistronic region can be translated independently. In the eukaryotic system the translation complex can initiate at only one site so that each mRNA can code for only one protein.

- Chaperones stabilize protein structures.
- Prion diseases, such as scrapie and bovine spongiform encephalopathy are associated with improper protein folding.
- The excretion of proteins involves a leader sequence that guides the peptide through the ER as it is synthesized by the ribosome. The leader sequence is cut off by a signal peptidase.

Additional questions for Chapter 22

1. What amino acid sequence would be coded for by the following mRNA sequence:

   AUGACAAAACACUCAUGA?

2. Circle the bases in the sequence in question 1 that could be mutated without changing the amino acid sequence.

3. Explain the abbreviation His-tRNA^{His} being sure to include why the His is used twice.

4. Describe the process of double selectivity in amino acid activation forming 'charged' tRNAs.

5. What does the 'S' stand for when describing the 30S, 50S, and 70S ribosomal units found in *E. coli*?

6. What is the role of the Shine–Dalgarno sequence on the mRNA?

7. What kind of a molecule is a ribosome?

8. What is the role of the IF3?

9. What does the 'f' stand for in fMet - tRNA$_f^{Met}$?

10. How many GTP molecules are involved in the initiation and translation of a 10-amino-acid peptide in a prokaryote?

11. In the eukaryotic system, when does the 60S ribosome subunit join with the 40S subunit?

12. What structural elements do typical leader sequence peptides have in common?

13. What is the difference between the *cis* region faces and the *trans* region faces of the Golgi vesicles?

14. What is the single-letter amino acid code for the following peptide:

    Glu–Leu–Val–Ile–Ser–Pro–Arg–Glu–Ser–Leu–Glu–Tyr?

15. What is the role of an anchor sequence?

16. Why would peptides destined for the mitochondrial matrix be kept partly unfolded by chaperones?

Chapter 23

....

Viruses and viroids

Chapter summary

This chapter describes the roles that viruses play both as disease-causing agents and as biochemical research tools. The mechanisms by which viruses infect host cells and the different modes of transferring their genetic material are presented with a specific emphasis on the retrovirus that causes AIDS. The role of oncogenic viruses and the occurrence of plant viroids are also highlighted.

Learning objectives

- ❑ The differences between a virus, a virion, and a viroid.

- ❑ The nature and composition of the nucleocapsid.

- ❑ The mechanisms by which a virus can gain entry into a host cell.

- ❑ The unique mechanism employed by bacteriophages to gain entry into the cell.

- ❑ The variety of nucleic acid materials that can carry the viral genetic information.

- ❑ The occurrence of and differences between (+) and (−) viral RNA, and the uniqueness of the situation involving retroviruses with reverse transcriptase activity.

- ❑ The mechanisms by which viruses are released from the cell.

- ❑ An overview of some of the specific characteristics of the action of the smallpox, polio. and influenza viruses.

- ❑ The role played by the integrase enzyme in the action of retroviruses.

- ❑ The mechanism of action of the drug azidothymidine (AZT) on the autoimmune deficiency virus (AIDS).

- ❑ The role of oncogenes and oncogenic retroviruses.

- ❑ The mechanism of the *E. coli* bacteriophage lambda which, after the integration of the viral DNA into the host chromosome, can remain inert until induced at a later date.

- ❑ The role of viroids in the infection of plants.

A walk through the chapter

Viruses

Viruses are small and have no ....(1). A virus particle is called a
...........(2). There are different viruses that infect ....(3), ...(4),
and ....(5). Their strategy is to get their genetic material into cells and
to use the ....(6) cell machinery for their own ....(7).
A protein shell surrounds the genome of the virus and is called the
..............................(8). The life cycle of a virus begins when it ....(9)
a cell and releases its genetic material which can be either RNA or DNA. Viruses can
gain entry into a cell via a ....(10) protein which is complementary to a
specific ....(11) on the host cell surface. The receptor–virus complex moves to a region of the membrane called a ....(12) which has the
appearance of a ....(13). The virion becomes engulfed into a vesicle
inside the cell, which ....(14) with an ....(15), and the clathrin
is recycled. Within the endosome, the virus dissociates from the receptor, sheds its
.......................(16), and finds its way into the cytoplasm. Certain viruses and
in particular the ....(17) viruses with a lipid membrane coat can ....(18)
with the external membrane of the host cell so that the contents of the virus enter the
host cell. Bacterial viruses known as ....(19), such as phage
...........(20), have a head that contains the viral DNA and a ....(21) that
serves as a needle to inject the genetic material of the virus into the cell.

Answers: (1) metabolism; (2) virion; (3) animals; (4) plants; (5) bacteria; (6) host; (7)
replication; (8) nucleocapsid; (9) infects; (10) coat; (11) receptor; (12) clathrin; (13) coated pit; (14)
fuses; (15) endosome; (16) nucloecapsid; (17) HIV; (18) fuse; (19) bacteriophages; (20) lambda;
(21) tail.

Complete the following

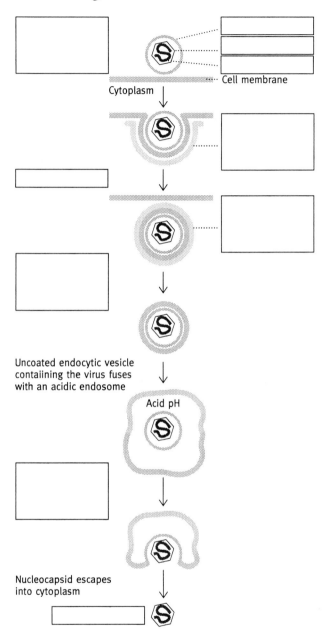

Cytoplasm

Cell membrane

Uncoated endocytic vesicle
contaiining the virus fuses
with an acidic endosome

Acid pH

Nucleocapsid escapes
into cytoplasm

Viral genetic material

Viral genetic material can come in many forms, as either ....(1) or ......(2) and in either ....(3) or ....(4) stranded forms. Viruses can function with RNA as the carrier of their genetic information because their genomes are ....(5), and the comparatively high level of ....(6) giving rise to a rapid rate of ....(7) may in some instances turn out to be an ....(8) by providing a way to escape immunological attack. Double-stranded viral DNA is transcribed by host RNA ....(9) producing ....(10), although some viruses carry their own polymerase enzymes. If the viral DNA arrives as a ....(11) strand, a complementary strand is formed by the host. Viruses with ....(12) stranded RNA carry their own ....(13) dependent RNA polymerase, which makes ....(14) from viral RNA. Viruses with single-stranded RNA occur in two forms, since the RNA strand might be an appropriate ....(15) strand termed (+) or it might be the ....(16) nonsense strand termed (–). The (+) strand is essentially ....(17) and this is ready to go for protein synthesis. The (–) strand of RNA must be ....(18) into the corresponding (+) RNA. The host cell does not have the machinery to perform this transformation. Some (+) strand RNA viruses, specifically the HIV, human ....(19) virus, which causes AIDS, acquired immune deficiency ....(20), are ....(21), in that the virion carries an enzyme that converts single-stranded .....(22) into double-stranded .....(23) which ....(24) into the host ....(25).

Answers: (1) RNA; (2) DNA; (3) single-; (4) double-; (5) small; (6) errors; (7) mutation; (8) advantage; (9) polymerase; (10) mRNA; (11) single; (12) double; (13) RNA-; (14) mRNA; (15) template; (16) complementary; (17) mRNA; (18) copied; (19) immunodeficiency; (20) syndrome; (21) retroviruses; (22) RNA; (23) DNA; (24) integrates; (25) chromosome.

Polio, smallpox, and influenza

Viruses can be released by ....(1) of the cell, as is the case in ....(2), while in bacterial cells that have been infected the enzyme ....(3) can be released to destroy the cell ....(4). Some viruses with lipid membrane ....(5) use the cell's own system of ....(6) in-volving the ER and Golgi to eventually ....(7) off the plasma membrane of the infected cell. The vaccinia virus which causes ....(8) and was used for ....(9) vaccination contains double-stranded ....(10) which finds its way into the ....(11) of the cell and is transcribed by host RNA polymerase into ....(12). The ....(13) virus is a naked virion with no ....(14) coat which attaches to specific ....(15) found only on the epithelial cells of humans and other primates. It arrives as a single-strand ....(16) which codes for, among other things, an RNA ....(17) that synthesizes (–) RNA which in turn is used as a ....(18) for more (+) RNA. The ....(19) used for this replication is a protein. The influenza virus is a ......(20) strand of RNA that contains eight sections. One of these, called ....(21),   interacts with the ......(22) blood cell protein glycophorin A on one of its carbohydrates composed of ....(23)   acid. The viral enzyme neuraminidase may be involved in a number of important processes such as the sneezing reflex. The virus carries its own RNA replicase which can make mRNA. The immune system responds to the hemagglutinin, but protection is ....(24) as this protein is mutated through ....(25) drift. This can be especially ....(26) when two different strains of a virus infect a cell and ....(27) their genetic material creating a ....(28) virus.

Answers: (1) lysis; (2) polio; (3) lysozyme; (4) wall; (5) envelopes; (6) translocation; (7) bud; (8) cowpox; (9) smallpox; (10) DNA; (11) nucleus; (12) mRNA; (13) polio; (14) membrane; (15) receptors; (16) mRNA; (17) replicase; (18) template; (19) primer; (20) (–); (21) hemagglutinin; (22) red; (23) neuramic; (24) lost; (25) antigenic; (26) troublesome; (27) mix; (28) new.

Retroviruses, oncogenes, bacteriophages, and viroids

Retroviruses are single-stranded ....(1) viruses, for example, the one causing AIDS, in which a ....(2) strand of RNA is packaged along with an enzyme with ....(3) transcriptase activity. This enzyme activity was not thought to be possible before this discovery. It was never thought that ......(4) could direct the synthesis of ....(5), the reverse of the norm. The virus uses a host cell tRNA as a ....(6) and makes a single-stranded ......(7) copy followed by a replication of this to make double-stranded DNA. The viral genome becomes ....(8) with an ....(9) enzyme into the host cell chromosome. The double-stranded DNA copy is called ....(10) DNA and has an LTR, long ....(11) repeat sequence, at each end. The proviral DNA is replicated along with host DNA for cell division. New virus particles are formed through the transcription of the proviral DNA into (+) RNA. The drug ....(12), azidothymidine, ...(13) the viral reverse transcriptase and the host cell DNA ....(14). There is hope that modified retroviruses may have beneficial applications in gene ....(15). Oncogenic retroviruses can cause ....(16), with the gene responsible being called an ....(17).   The cancer-producing ....(18)   oncogenes have similar counterparts in ....(19) cells, and may involve only a single base change. The viral gene is given the prefix v, while the corresponding ....(20) gene is denoted c, also called the ....(21). This gene often codes for a protein involved in an important cellular ....(22) mechanism, such that abnormalities lead to uncontrolled cell ....(23) and cancer. Bacteriophage ....(24) contains double-stranded DNA, with an elaborate set of genetic controls. The virulent, ...(25) path of events that follows virus infection involves ....(26) replication of the DNA and host cell ....(27) with the enzyme ....(28). A second pathway, known as the lysogenic route, involves integration of the viral DNA into the *E. coli* chromosome where it can remain ....(29) carrying the ....(30). The lysogenic cell can be induced at a later time to become virulent and ....(31) the phage. Viroids are small, naked RNA molecules that can infect ....(32). The RNA sequence lacks an open reading frame so that ....(33) do not appear to be coded. Much is not known about the way in which this class of infectious agents works.

Answers: (1) RNA; (2) (+); (3) reverse; (4) RNA; (5) DNA; (6) primer; (7) DNA; (8) integrated; (9) integrase; (10) proviral; (11) terminal; (12) AZT; (13) inhibits; (14) polymerase; (15) therapy; (16) tumours; (17) oncogene; (18) retroviral; (19) normal; (20) cellular; (21) protooncogene; (22) control; (23) division; (24) lambda; (25) lytic; (26) immediate; (27) lysis; (28) lysozyme; (29) inert; (30) prophage; (31) release; (32) plants; (33) proteins.

Complete the following

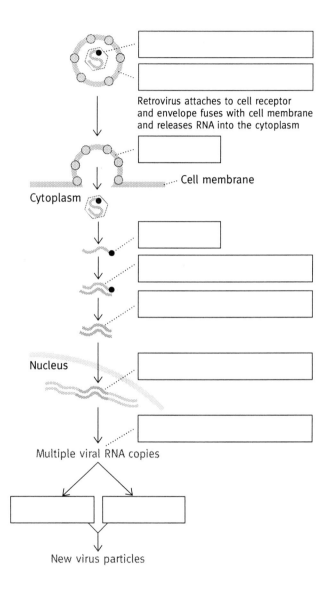

Retrovirus attaches to cell receptor and envelope fuses with cell membrane and releases RNA into the cytoplasm

Cell membrane

Cytoplasm

Nucleus

Multiple viral RNA copies

New virus particles

Review of problems from the end of Chapter 23

- Receptor-mediated endocytosis involves a very specific recognition event between a viral protein and a specific cell membrane receptor site.
- Negative (–) single-stranded RNA cannot be used to make protein directly since the information is contained in complementary code to be read by the tRNAs and

it runs in the opposite (3′–5′) direction along the chain. An RNA replicase enzyme is needed to generate the (+) RNA strand.

- The vaccinia virus can replicate directly in the cytoplasm since it carries its own RNA polymerase.
- The (+) single-stranded RNA could be transcribed directly into protein but, in the case of a retrovirus, DNA is made from this RNA template.
- The lipid membrane envelope comes from the host cell, starting from the ER and eventually budding off of the plasma membrane.
- Influenza epidemics often involve a recombination between different virus strains which eludes the natural immune protection response in a great number of people.
- The agglutination of red blood cells by influenza virus can be seen *in vitro* and is a consequence of the fact that the virus contains surface proteins that bind to sugar residues that are found in the exterior of red blood cells. The agglutination involves a crosslinking structure.
- The surface neuraminidase activity is surprising because the virus attaches to the host cell via the sugar residue's neuraminic acid. The logic behind this is not completely understood but may involve the mechanism of spreading the virus particles to other host cells.
- In the lysogenic pathway, the viral DNA is integrated into the *E. coli* cell chromosome and becomes dormant. In the lytic phase, the viral DNA is used to make new virus particles directly and the host cell is disrupted.
- Viroids are infectious RNA molecules that do not appear to code for proteins.

Additional questions for Chapter 23

1. What is the role of acid endosomes in the process of receptor-mediated endocytosis of viruses?
2. Approximately how many base pairs of DNA are present in a bacteriophage λ particle?
3. What two reactions are catalysed by the viral reverse transcriptase enzyme?
4. How is azidothymidine different from natural thymidine and how does this difference affect cell function?
5. What is a protooncogene?
6. What do oncogenes usually do?
7. How long does it typically take the coconut cadang-cadang viroid to kill a coconut palm?
8. What happens after a cell infected with bacteriophage λ DNA which has undergone the lysogenic pathway receives an induction signal?

Chapter 24

....

Gene cloning, recombinant DNA technology, genetic engineering

Chapter summary

This chapter describes the types of gene manipulations now possible as a result of the new recombinant DNA technology. The importance of restriction endonucleases, bacterial cloning vectors, and genomic libraries is presented and emphasized. The methods of the Sanger dideoxy DNA-sequencing technique, the polymerase chain reaction (PCR), and restriction fragment length polymorphism are highlighted in terms of their principles and applications.

Learning objectives

- ❏ The techniques for isolating individual genes.

- ❏ The techniques for determining and manipulating DNA sequences.

- ❏ The process of deducing protein sequences from DNA.

- ❏ The important characteristics of restriction endonucleases.

- ❏ The role of methylated bases in protecting the cell from its own endonucleases.

- ❏ The notion of cloning and the relationship between cDNA and the native strand.

- ❏ The role of bacterial cell lines in making copies of pieces of DNA and proteins.

- ❏ The role of cloning vectors and what kinds of things are combined to make recombinant DNA.

- ❏ The roles of 'sticky-ended' and 'blunt-ended' DNA in the ligation process.

- ❏ The nature of a genomic library using different *E. coli* strains and the process of screening using a hybridization probe.

- ❏ The rationale behind preparing a DNA probe by working backwards from the amino acid sequence and making allowances for redundancies.

- ❏ The usefulness of the polyA tail in human mRNA.

❑ The important role of the reverse transcriptase in making DNA from mRNA.

❑ The nature of plasmid DNA, and one of the versatile techniques for using pBR322 with combinations of antibiotics to select competent cells.

❑ The details and principles behind the use of the Sanger dideoxy method of sequencing DNA and the ability to read the sequence from the four lanes of an electrophoresis gel.

❑ The importance and application of the polymerase chain reaction (PCR) for amplifying a specific DNA segment.

❑ Examples of commercial applications of the process of mass-producing proteins using recombinant techniques.

❑ The principles behind the production of transgenic plants or animals with desired phenotypic characteristics.

❑ The use of restriction fragment length polymorphism (RFLP) techniques to identify the biological source of pieces of DNA.

A walk through the chapter

New techniques and restriction enzymes

DNA manipulation techniques make it possible to ....(1) individual genes, determine and manipulate their ....(2), and transfer them from one ....(3) to another. In addition, DNA codes can be read to deduce the ....(4) sequence. Previous to these advances the huge amount of nucleic acid material, distinguishable only by differences in ....(5), made these tasks very challenging. Restriction endonucleases, also called DNases, or, simply, ...(6) enzymes, from ....(7) cut the DNA at precise points in the sequence. The sequences which are cut have twofold symmetry, such that the ....(8) strand read in the $5' \rightarrow 3'$ direction has the ....(9) sequence. The biological function of these ....(10) appears to be one of ....(11) against invading foreign DNA. The cell guards against cutting its own DNA by adding a ....(12) group to one of the bases in all of the ....(13) sequences. Different strains of the bacteria have restriction ....(14) with different ....(15) sites.

Answers: (1) isolate; (2) sequences; (3) species; (4) protein; (5) sequence; (6) restriction; (7) bacteria; (8) complementary; (9) identical; (10) endonucleases; (11) protection; (12) methyl; (13) recognition; (14) enzymes; (15) recognition.

Cloning

Genomic cloning produces a piece of ....(1) identical in base sequence to the corresponding stretch of DNA. The isolation of cDNA, 'c' for ....(2), is a double-stranded DNA copy of a piece of ....(3) that differs from the native DNA in that it does not contain ....(4). Note that cloned eukaryotic genes ....(5) direct protein synthesis in ....(6), on their own, but this cDNA can be transcribed into ....(7) provided the appropriate ....(8) signals are in place. This modified transcript will direct protein ....(9) in bacteria. Cloning usually employs a bacterial cell for the ....(10) of a single piece of DNA and, since bacterial cells can quickly reproduce exact copies of themselves, a pure ....(11) of a selected cell ....(12) allows for virtually ....(13) copies of the specific piece of DNA. Cloning ....(14) are pieces of DNA made by combining pieces of DNA from one source with DNA that has the ability to ....(15) in *E. coli*. Lambda bacteriophage DNA can be ....(16) and a piece of human DNA with a size of about ....(17) kilobases inserted, creating a ....(18) phage. This recombinant piece of DNA is prepared from the necessary pieces of DNA by joining pieces with ....(19) which result from restriction enzyme cuts being made with ....(20) sequences that will automatically base pair with each other. Pieces with two identical sticky ends will join together. A ....(21) enzyme is used to seal the nicks. The vector can be packaged into the lambda phage assembly and can be used to ....(22) *E. coli* cells using a procedure that maximizes the chance that only ....(23) phage will enter a given cell. The strain of *E. coli* used must be a ....(24) that lacks its restriction enzyme. When many different pieces of DNA are used in the recombinant process, a genomic ....(25) is generated in which the entire fragmented ....(26) can be collectively contained in a ....(27) of different *E. coli* cells, one piece per cell. The phage-infected cells grow into ....(28), which can be distinguished from the uninfected cells.

Answers: (1) DNA; (2) complementary; (3) mRNA; (4) introns; (5) cannot; (6) bacteria; (7) mRNA; (8) transcription; (9) synthesis; (10) replication; (11) culture; (12) line; (13) unlimited; (14) vectors; (15) replicate; (16) modified; (17) 15–20; (18) recombinant; (19) 'sticky ends'; (20) overhang; (21) ligation; (22) infect; (23) one; (24) mutant; (25) library; (26) genome; (27) culture; (28) plaques.

Complete the following

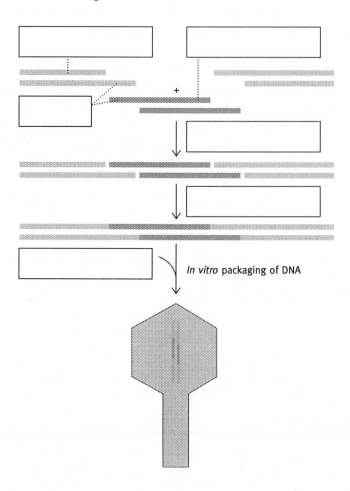

In vitro packaging of DNA

Screening a library

The plaques are then ....(1) using a ....(2) probe consisting of a short (~20 nucleotides) piece of DNA whose base sequence is complementary to a known sequence in the gene in question. The test involves ....(3) a bit of the phage from each ....(4) to a piece of DNA-absorbent membrane. After working this membrane through a series of steps, a ....(5) hybridization probe is used to identify which plaque has the piece of DNA of interest. The radioactivity serves in the ....(6) process since it can be detected with a piece of ....(7) film. Once the plaque of choice has been identified, unlimited numbers of the phage within this plaque can be grown by infecting additional *E. coli* cells. The ....(8) piece of DNA can be prepared synthetically by working ....(9)   from the amino acid ....(10) of the protein that is coded for. Allowances must be made for the redundancy of the genetic code and it may be necessary to make more than one probe. A human ....(11) library is made by isolating the ....(12) from cells expressing the gene in question; often this arrangement can be specifically induced. Almost all of the mRNA in human cells has a poly ....(13) tail while the other forms of RNA do not. This fact can be used as the basis of a purification technique by using a separation column containing a synthetic ....(14) ligand so that only the mRNA ....(15) to the column with the rest of the RNAs passing through. The mRNA is then eluted using solutions of low ionic strength. The mRNA is then copied into DNA using a purified viral reverse ....(16). With this the RNA is destroyed and the single-stranded DNA is converted to double-stranded DNA by an exonuclease-free DNA ....(17). The DNA is then made into recombinant ....(18), often using blunt-end ligation, and ....(19) into bacteria.

Answers: (1) screened; (2) hybridization; (3) transferring; (4) plaque; (5) radioactive; (6) identification; (7) photographic; (8) probe; (9) backwards; (10) sequence; (11) cDNA; (12) mRNA; (13) A; (14) oligo-dT; (15) stick; (16) transcriptase; (17) polymerase; (18) vectors; (19) cloned.

Plasmids

Once inserted into the phage, the DNA can be amplified and recovered easily. Frequently, the DNA, once obtained, is inserted into a bacterial ....(1), which is much smaller and easier to handle. Depending on the size of the DNA and the goals of the experiment, either ....(2) plasmids or an ....(3) vector is used. The expression vectors have the appropriate transcriptional and translational signals needed for the *E. coli* to produce the ....(4). Plasmids are ....(5) pieces of DNA that can be duplicated inside the cell. The cells can be treated and made ....(6) to take up the plasmid from the surrounding buffer. The plasmid can be made ....(7) by opening the plasmid, inserting the cDNA, and reclosing the plasmid. The plasmids are put into a group of *E. coli* cells in such a way that only a small population of the cells become infected and the possibility of two plasmids being taken up by one cell is minimized. The subpopulation of the cells that have the plasmid can be selected by using a technique like the one engineered into the ....(8) plasmid, pBR322, which has genes for ....(9) and ....(10) resistance, with different ....(11) enzyme sites within each of these genes. Thus, depending on where the extra piece of DNA is inserted in the plasmid, the resistance of the cell toward one of these ....(12) can ....(13). Native or wild *E. coli* are inhibited by both antibiotics, while bacteria with the plasmid will be ....(14) to both and cells with the ....(15) that contains the inserted cDNA will be resistant to only the ....(16) whose gene was not interrupted. In this way the appropriate cells can be ....(17).

Answers: (1) plasmid; (2) sequencing; (3) expression; (4) protein; (5) circular; (6) competent; (7) recombinant; (8) cloning; (9) ampicillin; (10) tetracycline; (11) restriction; (12) antibiotics; (13) change; (14) resistant; (15) plasmid; (16) one; (17) selected.

Sequencing DNA

Two methods used to sequence a cloned piece of DNA are the direct chemical method of ....(1) and the other, based on enzymatic replication of the DNA, called the Sanger ....(2) method. The Sanger method involves copying the DNA to be sequenced using the DNA ....(3) enzyme. This reaction requires the four deoxynucleoside ....(4), a piece of single-stranded ....(5) to be copied, and a ....(6) that is hybridized to the start site. In addition, a small amount of a ....(7) derivative of one of the nucleoside triphosphates, ...(8), is included. The 'N' in NTP is used to mean any one of the four nucleic acid bases. The course of the reaction is followed by the incorporation of a ....(9) element into the molecule. An ....(10) method is used to ....(11) the DNA molecules after they are made and is sensitive enough to separate pieces differing in length by only one nucleotide. The chain grows as the polymerase adds a nucleoside to the ....(12) of the previous base. The novel aspect of this technique that makes it possible to determine the sequence is that, when a dideoxy NTP is added, it ....(13) the 3'-OH, and the chain is ....(14). Note that the dideoxy NTP can be added because it does have a ......(15) phosphate but that the next base cannot and so one knows that the last nuclotide in the sequence is the dideoxy one. By only using a very ....(16) amount of a single dideoxy NTP in each run, most of the chains that are synthesized will have a ....(17) base added, but a fraction will be ....(18), with the ....(19) base as its terminal nucleotide. The terminated chains will be detected as separate ....(20) on an electrophoresis gel. The procedure is repeated for each of the four ....(21) NTPs used, A, G, C, and T. All four incubations can be carried out simultaneously and the products run side-by-side on a gel producing a ....(22) ladder. The sequence is read off from the bottom of the gel upwards and represents the sequence of the ....(23) chain, the partner of the template. This gives the ....(24) of the newly synthesized complementary strand of DNA in the 5' → 3' direction. The sequence of the original strand is complementary and antiparallel to this newly synthesized strand. About ....(25) bases can be determined in each gel so that a strategic overlapping of pieces is used to sequence very large pieces of DNA. For determination of the coded ....(26) acid sequence, the correct ....(27) frame must be found.

Answers: (1) Maxam–Gilbert; (2) dideoxy method; (3) polymerase; (4) triphosphates; (5) DNA; (6) primer; (7) dideoxy; (8) ddNTPs; (9) radioactive; (10) electrophoretic; (11) separate; (12) 3′-OH (13) lacks; (14) terminated; (15) 5′; (16) small; (17) normal; (18) terminated; (19) specific; (20) bands; (21) dideoxy; (22) sequencing; (23) copy; (24) sequence; (25) 200–300; (26) amino; (27) reading.

Complete the following

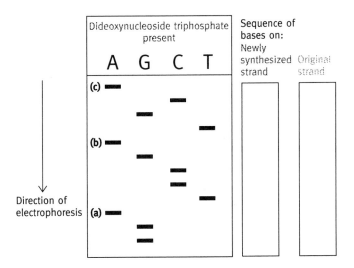

Polymerase chain reaction

The polymerase chain reaction, ...(1), is a technique for ....(2) a specific DNA segment; it represents a major advance because only a small amount of DNA is needed initially. The technique involves taking a piece of double-stranded DNA and separating the strands by ....(3); a selected portion of the DNA is then copied, and the process is ....(4) again and again. The section of the longer piece of DNA to be copied is selected by using ....(5) corresponding to the limits of the section to be amplified. Different primers are needed for each end so that something must be known about the base ....(6) of the DNA to which the primers will ....(7). The degree of amplification is ....(8) since each of the original pieces and each previous copy are copied in each cycle, giving an ....(9) increase. One of the discoveries that made this practical was that of ....(10) DNA polymerases from thermophilic ....(11).

Answers: (1) PCR; (2) amplifying; (3) heating; (4) repeated; (5) primers; (6) sequence; (7) hybridize; (8) enormous; (9) exponential; (10) heat-stable; (11) bacteria.

Applications

Applications of these techniques include using a bacterial ....(1)   vector to produce ....(2) from an isolated piece of cDNA. This has found widespread use with human proteins of ....(3) value that would be very difficult to obtain via ....(4) directly from natural sources. One example of this class is human ....(5). Another advantage of the use of a bacterial expression system over isolation from human sources is that production in *E. coli* ....(6) the danger of such a protein being ....(7)   with a human ....(8)   agent. Often, the bacteria will produce so much of a given protein that it precipitates out as an ....(9) body. Proper protein folding is often a ....(10)   in the procedure. The need for ....(11), which bacterial cells do not do, can be overcome by using ....(12) cell systems such as insects.

Transgenesis involves the insertion of ....(13)   into plants or animals to give different ....(14)   characteristics. One aspect of this could involve gene ....(15), although methods for controlling the specific point of insertion are still only being developed. In ....(16), cloning vectors and gun-type instruments have had success in giving plants ....(17)   to herbicides. Transgenic animals have been obtained and used for many purposes. In one example the desired protein was expressed in the animal's milk. Genes can be analysed for certain ....(18) with great speed and precision by these new techniques. For example, hybridization ....(19)   can be made to detect gene ....(20)   on extremely small samples such as those that could be obtained from a living ....(21). The technique used to detect such sequences is a ....(22)   blot analysis; this involves cutting the DNA with one or more restriction enzymes, and, after separating the fragments on an electrophoresis gel, ...(23)   the fragments to a membrane and probing the membrane with a specific radioactive hybridization probe. The pattern of bands is characteristic for certain diseases. With the use of RFLP, restriction fragment length ....(24)   analysis, repetitive human DNA sequences located throughout the genome give patterns that are ....(25) to each ....(26).   RFLP   can   thus   be   used   as   a   method   of   DNA ....(27).   An even newer, more sensitive application of this type of analysis uses the ....(28) reaction also.

Answers: (1) expression; (2) protein; (3) therapeutic; (4) isolation; (5) insulin; (6) eliminates; (7) contaminated; (8) infectious; (9) inclusion; (10) complication; (11) glycosylation; (12) animal; (13) genes; (14) phenotypic; (15) therapy; (16) plants; (17) resistance; (18) sequences; (19) probes; (20) abnormalities; (21) fetus; (22) Southern; (23) transferring; (24) polymorphism; (25) unique; (26) individual; (27) fingerprinting; (28) PCR.

Review of problems from the end of Chapter 24

- A restriction endonuclease is a DNase with very exacting specificity for the sequence of bases around where the cut is made.

- Bases of this sequence in native DNA are methylated and thus protected from attack by the cell's own endonuclease.

- Two sticky ends will automatically base pair if the sequences are complementary.

- cDNA is made from mRNA that has had its intron sequences removed. A genomic clone is a copy of the DNA with the introns and the promoter sequences present.

- A DNA library is a collection of cell lines (usually *E. coli*) in which each of the lines has a different piece of DNA from another source. The pieces are initially cut so that many combinations of starting and stopping points for a given length are made. The λ bacteriophage system is so useful because of its ability to self-associate, package the DNA, and deliver it to the strains of *E. coli* cells.

- The process of isolating a particular clone from a genomic library is known as screening. After a group of cells have been infected, the first step is to select only the cells that incorporated the foreign DNA-containing phage and the second step is to select the individual cell that has the desired piece of DNA by examining the phage that is produced from this cell. This second selection is made specific by using a radioactively labelled piece of DNA, known as a hybridization probe, that will bind to the complementary desired DNA sequence and make it detectable due to the ability of the radioactivity to mark its location via an X-ray film.

- A dideoxynucleoside triphosphate lacks sugar hydroxyls on both the C-2' and C-3' carbons. When used in the Sanger sequencing method those compounds cause periodic terminations of the polymerase reaction. A correlation can be made between the size of the piece caused by such a termination event and the position of the base in the sequence of the DNA. For example, ddCTP will terminate the chain at a position where the C base occurs.

- The polymerase chain reaction allows the amplified reproduction of a specific section of DNA. In order to apply this technique, primer pieces of DNA for copying the desired section in both directions are needed.

- Expression vectors are used for preparing bacteria strains that can express proteins.

- The pattern of restriction sites is characteristic of the individual and its genetic makeup.

Additional questions for Chapter 24

1. How are restriction enzymes such as *Eco*RI and *Bam*Hl named?

2. Explain why unlimited numbers of DNA pieces can be obtained from a bacterio-phage library.

3. Which of the following DNA sequences is consistent with a portion of the template strand for the Sanger sequencing gel shown:

    (a) 5′ TCTGACCAG 3′;

    (b) 5′ GAGGGTCAG 3′;

    (c) 5′ CTCATTGAG 3′;

    (d) 5′ GACCAGTCT 3′;

    (e) 5′ TGACTGGCT 3′?

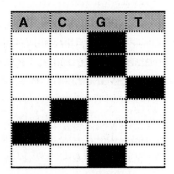

4. What advantage is there to using the DNA polymerase from the thermophilic bacteria in the PCR reaction rather than just any polymerase?

5. What two conversions need to be performed when reading a Sanger sequencing gel?

6. If four single strands of DNA are run through five cycles of the PCR reaction, how many strands are made?

7. What is a Southern blot analysis?

8. What is a genetic linkage analysis?

9. What is the logic behind the use of RFLP as a fingerprinting technique?

Chapter 25

The immune system

Chapter summary

This chapter describes one of the main mechanisms whereby an organism can protect itself from foreign, pathogenic agents such as bacteria and viruses. The two main protective mechanisms involving antibodies and cytotoxic killer cells are described. The importance of suppressing the autoimmune response is stressed. The general structure of the five main immunoglobulins is outlined. The mechanism of activation involving the major histocompatibility complex (MHC) is presented, as well as the roles of helper T cells and CD4 and CD8 molecules. Important references to the effect of the AIDS virus are highlighted.

Learning objectives

❏ The major function of the immune system.

❏ The nature of the problem that causes autoimmune diseases and some examples of these.

❏ The two main protective mechanisms involving antibodies and cytotoxic killer cells.

❏ The functions of B cells, helper T cells, and killer T cells.

❏ The mechanism of stimulating B cells with helper T cells.

❏ The importance of the facts that each B cell can produce only one antibody and that each T cell responds to only one antigen.

❏ A proposed mechanism for the suppression of the autoimmune response.

❏ The role of the protein perforin in the process of apoptosis.

❏ The general structure of immunoglobulins.

❏ The five major classes of immunoglobulin and some of their functional differences.

❏ The method by which the cell achieves antibody diversity.

❏ The occurrence of allelic exclusion and affinity maturation.

❏ The role of the major histocompatibility complex (MHC) in the process of B cell activation.

❑ The role of the T cell receptors (TCRs) on the helper T cells that recognize the MHC–antigen complex.

❑ The role of the antigen-presenting cells (APCs) which also display the class II MHC receptors.

❑ The role of the CD4 protein in general and with specific reference to the AIDS virus.

❑ The role of cytokines released when a helper T cell encounters a B cell displaying the same MHC–antigen complex as that originally presented to the T cell by the APC.

❑ The role of memory B cells.

❑ The mechanism by which killer T cells recognize and kill body cells that have become abnormal, and the role played in this process by the CD8 glycoprotein.

❑ The reason why it might be advantageous that MHC molecules are different in each individual.

A walk through the chapter

The nature of the immune system

The immune system protects the body from ....(1) organisms such as bacteria and viruses. One of the most important principles in the immune system is the need to avoid reacting against the proteins and other molecules of the body itself. Problems associated with an error in this system result in ....(2) diseases. Some examples are given in the following table.

Autoimmune disease	Effects
..............................(3)	Destroys the acetylcholine receptors of muscle
..............................(4)	Produces antibodies that crossreact with a protein component of the heart
Insulin-dependent diabetes	....(5)

The two protective mechanisms of the immune system are the production of ....(6) (also known as the humoral response) and ....(7) immunity. Antibodies are proteins that bind to ....(8), which are molecules that produce an immune response. In cell-mediated immunity, ...(9) killer cells destroy infecting cells by making direct contact with them. Antibodies are produced by ....(10), called plasma cells, that come from stimulated B cells, while killer cells are T lymphocytes, T for ....(11) derived.

All lymphocytes originate in the ....(12) marrow from ....(13). B cells become stimulated when they meet an ....(14) and are contacted by a ....(15) cell that has met the ....(16) antigen. The ....(17) of lymphocytes is antigen-independent, while ....(18) of the mature lymphocyte is antigen- ....(19).

Cell type	Function
.........................(20)	Produce antibodies after activation and maturation into plasma cells
.........................(21)	Are involved in the activation and maturation of B cells
.........................(22)	Bind to host cells dispaying a foreign antigen and kill them by perforating their membrane or inducing apoptosis

Answers: (1) pathogenic; (2) autoimmune; (3) Myasthenia gravis; (4) Rheumatic fever; (5) Destroys pancreatic cells; (6) antibodies; (7) cell-mediated; (8) antigens; (9) cytotoxic; (10) lymphocytes; (11) thymus-; (12) bone; (13) stem cells; (14) antigen; (15) helper T; (16) same; (17) maturation; (18) differentiation; (19) dependent; (20) B cells; (21) Helper T cells; (22) Killer T cells.

Complete the following

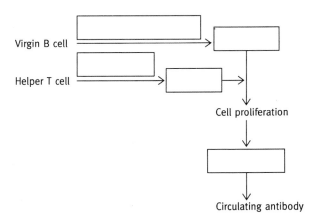

Virgin B cell

Helper T cell

Cell proliferation

Circulating antibody

Overview of strategy

Of the vast numbers of B and T cells produced, each B cell can produce only one specific ....(1) and each T cell responds to only a single ....(2). One mechanism proposed to explain how the body might suppress the ....(3) response is that B cells as they mature are ....(4) to a large proportion of the components of ....(5), and any ....(6) that might produce ....(7) to these are eliminated. The process involves the immature B cell ....(8) its antibody molecules on its ....(9). If an immune response occurs, that B cell is ....(10).

T cells also display antigen-specific ....(11) during their maturation in the thymus and undergo a similar ....(12) for reaction against self. This theory has come to be known as ....(13) selection theory. An exception to this is that some of the lipopolysaccharides of Gram-negative ....(14) cell walls are able to stimulate B cells without needing helper T cells. Part of the mechanism by which killer T cells destroy cells carrying foreign antigens involves the protein ....(15) which makes the membrane of the infected cell ....(16) and induces the programmed self-destruction of the cell called ....(17).

Answers: (1) antibody; (2) antigen; (3) autoimmune; (4) exposed; (5) self-antigens; (6) B cells; (7) antibodies; (8) displaying; (9) surface; (10) eliminated; (11) receptors; (12) screening; (13) clonal; (14) bacterial; (15) perforin; (16) leaky; (17) apoptosis.

Immunoglobulins

Antibodies are proteins of the class called ....(1), Ig. IgG is a Y-shaped molecule with two identical ....(2), L, polypeptide chains and two identical ....(3), H, chains held together by disulfide bonds. There are also ....(4) regions of the chains which form the ....(5) binding sites, one on each arm, which are identical. The specific part of the antigen that is recognized by the antibody is called an ....(6). Antigens and antibodies combine to form networks that can activate the ....(7) system, a group of proteins in the ....(8) that forms a cascade mechanism that results in ....(9) of the attacked cell's membrane. Alternatively, the antibody-coated cell can be ....(10) by phagocytic ....(11) via receptor-mediated processes. The ....(12) major classes of immunoglobulins differ from one another in the constant regions of their ....(13), H chains.

Class of immunoglobulin	Functions/properties
...................(14)	First antibody produced; polymeric form with 10 antigen combining sites
...................(15)	Results from repeated challenges by the same antigen.
...................(16)	Is important in the first line of defence because it is transported through epithelial cell membranes and thus is active in external mucous membranes
...................(17)	Sensitizes mast cells and is involved in histamine release and allergic symptoms. It has a role in the synthesis of PAF (platelet-activating factor) by certain sensitized individuals
...................(18)	Functions are unknown

It is interesting that B cells can switch from the production of one type of ....(19) to another, while still showing a specificity to the same antigen. Antibody diversity is *not* achieved by the cell having a separate ....(20) for each possible antibody. The process is based on the fact that the ....(21) region of the L chain is identical in all Ig molecules, while there are about 300 sections of DNA that can code for the ....(22) region (V) and four different sections that code for the ....(23)   section (J). Recombination events join one of the V sections with one of the J sections by a process in which the ....(24) of the piece is random. The exact DNA sequences of ligation are imprecise so that about ......(25) L chain variants are formed. The H chain genes undergo a similar recombination by random selection so there are many possible ....(26) produced, while only one type is produced from each individual B cell. A process known as ....(27) exclusion ensures that only one of the two sets of chromosomes expresses the assembled ....(28) gene. There is also a process known as affinity ....(29)   based on the fact that during B cell multiplication there is mutation in the V region, and the ....(30) that produce the most effective antibodies are selected and multiplied.

Answers: (1) immunoglobulins; (2) light; (3) heavy; (4) variable; (5) antigen-; (6) epitope; (7) complement; (8) blood; (9) perforation; (10) engulfed; (11) leukocytes; (12) five; (13) heavy; (14) IgM; (15) IgG; (16) IgA; (17) IgE; (18) IgD; (19) immunoglobulin; (20) gene; (21) constant; (22) variable; (23) joining; (24) selection; (25) 3000; (26) antibodies; (27) allelic; (28) immunoglobulin; (29) maturation; (30) B cells.

Complete the following

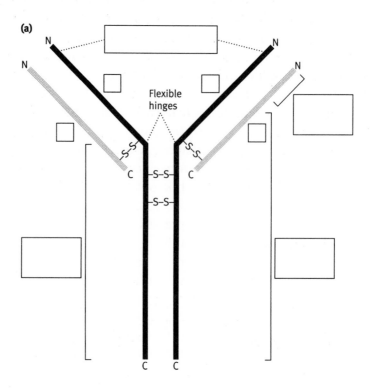

The activation process

B cells become activated when an ....(1) binds to the ....(2) displayed on the B cell membrane. This event causes the B cell to ....(3) the antigen and process it into small ....(4), which are displayed once again on the surface of cells in the binding sites of MHC (major ....(5) complex) molecules. This is what is recognized by the ....(6) cells with receptors called ....(7), T cell receptors, specific for the ....(8) complex. The helper T cells recognize antigens after the antigen has been engulfed by a special ....(9) called the APC, antigen-.......................(10) cell, which displays the antigen with the ....(11) MHC receptors. This process is facilitated by a ....(12) (cluster of differentiation 4) protein, which happens to be the receptor to which the AIDS virus attaches. Helper T cells become activated when they combine with the appropriate ....(13) assembly displayed on an ....(14). When an activated helper T cell encounters a ....(15) which is displaying the same MHC–antigen complex that was ....(16) presented to the T cell by the APC, the T cell binds and releases ....(17), which cause the B cells to

...................(18) producing ....(19) cells that produce and release
.........................(20). A proportion of the B cells that were activated become
..................(21) cells, which can circulate for years and can form the basis for a
rapid response to a repeat infection. The killer T cells have ....(22)
that recognize antigens on ....(23) MHC molecules via a surface glycoprotein
........(24) that interacts with the MHC class I. Killer T cells recognize and kill body
cells that have become ....(25). Viral infected cells will display ....(26)
of the viral protein on the cell surface with class I .....(27) molecules.

Answers: (1) antigen; (2) antibody; (3) internalize; (4) fragments; (5) histocompatibility; (6)
helper T; (7) TCR; (8) MHC–antigen; (9) macrophage; (10) presenting; (11) class II; (12) CD4;
(13) MHC–antigen; (14) APC; (15) B cell; (16) originally; (17) cytokines; (18) multiply;
(19) plasma; (20) antibodies; (21) memory; (22) receptors; (23) class I; (24) CD8; (25) abnormal;
(26) pieces; (27) MHC.

Complete the following

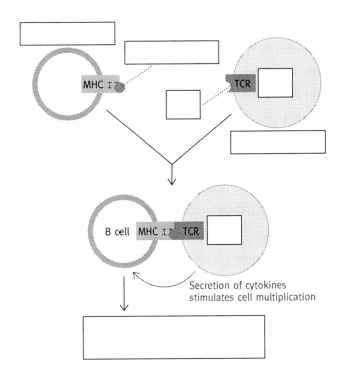

Foreign human cells

The body's immune system fiercely rejects ....(1) human cells because
the MHC molecules from one individual are different from those of another. The
genes are said to be ....(2) among the population, having many

...................(3), or variants. Any differences in MHC molecules will be ....(4); thus a foreign MHC molecule would be viewed by the immune system cells as the body's own MHC molecule complexed to a ....(5) antigen. This difference in MHC molecules helps to reduce the chance that a ....(6) could sweep through the whole ....(7) without being recognized as foreign.

Answers: (1) foreign; (2) polymorphic; (3) alleles; (4) antigenic; (5) foreign; (6) disease; (7) population.

Review of problems from the end of Chapter 25

- All of the type G immunoglobulins have two 'heavy' chains and two light polypeptide chains, covalently linked by disulfide bonds. Each molecule can also be divided into a constant region, which is similar among different individual IgGs, and a variable region which shows a much greater diversity between individual molecules as its name suggests. The overall structure of the molecule is that of the letter Y, with the antigen-binding domains on the variable region at the ends of the forks on the Y.

- There are numerous sections on the DNA, including 300 of the V type (variable) and 4 of the J type (joining) which can be recombined in a number of variations to include one V and one J segment. The process by which this occurs involves the arrangement of the segments of the gene to produce a primary RNA transcript which is further spliced to give the final mRNA for the immunoglobulin light chain.

-

Cell type	Function
B cells	Produce antibodies after activation and maturation into plasma cell
Helper T cells	Are involved in the activation and maturation of B cells
Killer T cells	Bind to host cells displaying a foreign antigen and kill them by perforating their membrane or inducing apoptosis.

- Antigen-presenting cells recognize, bind, internalize, process, and mark (with class II MHC) antigens, which are then presented on the outside of the cell.

- Class I MHC molecules are present on most cells of the body. Killer T cells interact only with host cells displaying a foreign antigen on their class I MHC molecules. Thus, the same class I MHC molecules are displayed on the host cell and the cytotoxic T cell.

- The selection of which 'clone' of B cells to grow and reproduce is dependent on interaction with the antigen.

- If a B cell interacts with an antigen in the bone marrow or if a T cell interacts with an antigen in the thymus, that cell is eliminated.
- The immunoglobulin is made without an anchoring polypeptide sequence and so does not need to be attached to the membrane.
- After the initial activation of the B cell, further mutations in the variable site of the immunoglobulin work in concert with an antigen-dependent selection system to produce the best possible antibody.
- Helper T cells have CD4 and cytotoxic T cells have CD8. CD4 binds to the class II MHC and CD8 binds to class I MHC. The CD4 protein is the receptor by which the AIDS virus infects the helper T cells.

Additional questions for Chapter 25

1. What is an antigen?
2. What is another common term for macrophages?
3. How are stem cells different from cells committed to forming one specific cell type?
4. What process involves erythropoietin?
5. What is the meaning behind the term virgin B cell?
6. What types of molecules are able to stimulate the relevant virgin B cell to produce its antibody without needing a helper T cell signal?
7. What is an epitope in terms of antibody binding?
8. Can more than one antibody bind to a single antigen molecule?
9. What are memory cells?
10. What are cytokines and what role do they play in the immune response?

Chapter 26

....

Chemical signalling in the body

Chapter summary

This chapter describes the major components involved in the extrinsic control mechanisms of chemical signalling that occurs in the body. The importance of the connection between uncoordinated cell activity and cancer is stressed. The classifications of signal molecules on the basis of chemical structure and mode of action are useful for viewing the different systems in context. The major pathways highlighted include steroid hormone action, G-protein-mediated events, the cAMP second messenger pathway, the action of nitric oxide, the phosphatidylinositol cascade, the receptors with tyrosine kinase activity, the action of interferons, and pathways involving voltage-gated and ligand-gated membrane channels. The occurrence and role of oncogenes and their relationship to cancer are described and contrasted with the normal state.

Learning objectives

- ❏ The connection between uncoordinated cell activity and cancer.

- ❏ The classification of control mechanisms according to whether or not they involve the regulation of gene expression.

- ❏ The classification (and some specific examples) of the types of signal molecules on the basis of chemical structural type and water versus lipid solubility.

- ❏ Some examples of the endocrine hormones known as the 'classical' signalling molecules.

- ❏ The convention of naming growth factors for the cell type associated with their function (cf. PDGF, EGF, and CSF) and the distinction between paracine and autocrine action.

- ❏ The role of neurotransmitters in delivering a signal to a target cell.

- ❏ The roles of the endocrine gland and tropic hormones in the action of the endocrine system.

- ❏ The importance of the principle of reversibility in regard to control mechanisms.

❑ The roles of zinc fingers, chaperones, and palindromic DNA sequences in the mechanism of steroid hormone action.

❑ The importance of the role of protein phosphorylation in an overview of receptor-mediated responses.

❑ The reason why the cAMP second messenger pathway can be involved in so many different effects in different cells.

❑ The subunit structure of the regulatory G proteins and the roles played by GTP, GDP, and the hydrolysis reaction.

❑ The reaction catalysed by the enzyme adenylate cyclase.

❑ The occurrence of both G_s stimulatory proteins and G_i inhibitory proteins.

❑ The mechanism by which cAMP affects gene transcription involving CRE, CREB, PKA, and leucine zipper type complexes

❑ The role of cGMP as a second messenger.

❑ The occurrence and action of nitric oxide (NO) as a second messenger.

❑ The details of the phosphatidylinositol (PI) cascade, and the role of phospholipase C (PLC) and the reaction that this enzyme catalyses.

❑ The roles of Ca^{2+} and diacylglycerol (DAG) as part of the PI cascade.

❑ The mechanism of action of phorbol esters in promoting tumour growth.

❑ The role of calmodulin in pathways that involve Ca^{2+} as a regulator.

❑ The occurrence of different combinations of hormones with second messenger mechanisms in different cell types.

❑ The occurrence of receptors that can self-phosphorylate with tyrosine kinase activity.

❑ The importance of the Ras protein mechanism, and the roles of the GRB, Raf, and SOS proteins on the mechanism of this pathway.

❑ The classification of certain proteins as protooncogene products, and their relationship to cancer-producing proteins coded for by retroviruses.

❑ An understanding of some of the mechanisms that can cause the conversion of protooncogenes to oncogenes.

❑ The importance of the p53 tumour suppressor gene.

❑ The mechanism of action of interferons and the roles of ISREs and JAK kinases.

❑ The mechanism involving ion channels and electrical membrane potentials whereby a neuronal signal conveys its effect.

❑ The role of ligand-gated channels in vision and the roles of rhodopsin, Ca^{2+}, and cGMP in this process.

A walk through the chapter

Classification of signal molecules

Extrinsic control factors come from ....(1) the cell and are important for the integration of various cell activities. Uncoordinated cell activity can be a form of ....(2). Cells that receive the message are called ....(3) cells and have a ....(4) to which the signal molecule binds. The cellular activities that are controlled by extrinsic factors can be divided into two groups—those that do and those that do not involve the regulation of ....(5) expression.

Examples of activities that do not involve the regulation of gene expression
Enzyme activities in fat and carbohydrate metabolism
Activation of voluntary striated muscle contraction by acetylcholine
Opening of ligand-gated pores in nerve cell membranes triggered by a nerve impulse
Examples of activities that involve the regulation of gene expression
Almost everything else

Transcription factors control gene ....(6) predominantly at the level of ....(7). The mechanism by which extrinsic control factors exert effects on the ....(8) of specific genes often involves membrane-bound ....(9), although a small number of lipid-soluble signalling molecules pass through the membrane and encounter ....(10) receptors.

Chemical type of signal molecule	Example	Soluble in
Proteins	Insulin	Water
Large peptides	Glucagon	................(11)
Small peptides	Vasopressin	................(12)
..........................(13)	Sex hormones	Lipid
Eicosanoids	..........................(14)	................(15)
Catecholamines	..........................(16)	Water
Small neutral molecules	Nitric oxide (NO)	................(17)

A classification in terms of biological ....(18) involves three groups—endocrine ....(19), growth factors and ....(20), and neurotransmitters—with ....(21) oxide being in a class of its own. The endocrine hormones are the 'classical' signalling molecules and have been known for a long time. They are produced in large amounts by specialized cells that secrete them directly into the circulation.

Hormone	Secreting organ	Target tissue	Function
................(22)	Pancreas	Liver, muscles	Stimulates gluconeogenesis, lipogenesis, protein synthesis
................(23)	Pancreas	Liver	Stimulates glycogen break-down, lipolysis
................(24)	Gonads	Reproductive organs and secondary sex organs	Promotes maturation and function in sex organs
Calcitonin	..................(25)	Bone, kidney	Inhibits reabsorption of Ca^{2+}
Somatotropin	Anterior pituitary	....................(26)	Stimulates synthesis of insulin-like growth factors IGFI and IGFII
IGFI	Liver	Liver, bone	..............................(27)

Answers: (1) outside; (2) cancer; (3) target; (4) receptor; (5) gene; (6) expression; (7) initiation; (8) transcription; (9) receptors; (10) intracellular; (11) Water; (12) Water; (13) Steroids; (14) Prostaglandins; (15) Lipid; (16) Epinephrine; (17) Lipid; (18) function; (19) hormones; (20) cytokines; (21) nitric; (22) Insulin; (23) Glucagon; (24) Testosterone; (25) Thyroid; (26) Liver; (27) Stimulates growth.

Complete the following

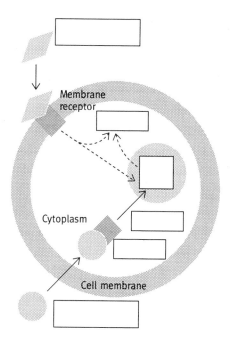

Comparison of signal molecules

Growth factors, also called developmental ....(1) factors, are unlike endocrine hormones in that they are secreted by cells that are not ....(2)
for this purpose. They often are named for the cell type associated with their
function (cf. PDGF, ...(3) growth factor, and EGF, epidermal growth factor). Other examples include CSF, ...(4)
factor, and ....(5), named after leukocytes. Growth factors involved
in the immune response are called ....(6). These factors may have quite
varied effects on cells. They combine with external cell ....(7)   and
indirectly produce effects on the initiation of specific gene ....(8).   In
contrast to the endocrine hormones, most growth factors diffuse only a ....(9)
distance and act ....(10)   on other cells (paracrine), while some are self-
stimulatory in that they act on the cells secreting them (autocrine). A signal delivered
via a ....(11) goes quickly and precisely to the target cell. Two
examples are the involuntary, ...(12) nervous system, which
causes the secretion of ....(13) in fat cell deposits, and the motor
neurones, which trigger ....(14) contraction via ....(15)
release. Endocrine hormones are released in response to ....(16)
hormones that control the release of other hormones from the anterior part of the
.......................(17)   gland located in the brain. Hypothalamic hormones are
alternatively called ....(18) factors; they cause target
endocrine glands to release their hormones. The anterior pituitary hormones bring
forth hormones from other glands and are thus called tropic hormones or tropins.
There is some degree of ....(19) control of the endocrine system, and
not all endocrine glands are controlled by this mechanism (cf. insulin and glucagon).
Growth factor release is an area under active research, and the mechanisms are
numerous. Platelets release ....(20),   which stimulates cell division and
..............(21) in response to epithelial cell damage of blood ....(22).   The
essence of control is reversibility; thus, released signal molecules must be removed
or else an initial signal would last indefinitely. The hydrolysis of acetylcholine by a
specific ....(23) at the nerve synapses is a good example.

Answers: (1) regulatory; (2) specialized; (3) platelet-derived; (4) colony-stimulating; (5) inter-
leukins; (6) cytokines; (7) receptors; (8) transcription; (9) short; (10) locally; (11) neurotransmitter;
(12) sympathetic; (13) epinepherine; (14) muscle; (15) acetylcholine; (16) hypothalamic; (17) pituitary; (18) hormone-releasing; (19) cerebral; (20) PDFG; (21) repair; (22) vessels; (23) esterase.

Comparison of pathways

Steroid hormone receptors have ....(1) finger DNA binding domains that exist in association with ....(2), heat shock proteins, which are also ....(3). In the case of the glucocorticoid receptor, binding of the ligand causes the complex between the receptor and heat shock protein to ....(4). The receptor is then transported into the ....(5) where it combines as a dimer with DNA at the specific glucocorticoid-responsive ....(6). This element of the DNA is a ....(7), and each half combines with the two zinc fingers on the receptor dimer. In ....(8) receptor-mediated responses, binding of the ligand either ....(9) an internal domain of the receptor or causes a ....(10) of receptors followed by, or associated with, activation of internal domains. The predominant theme in the overview of membrane receptor-mediated responses is protein ....(11) by protein kinases using ATP, along with ....(12) of the phosphoproteins by ....(13). Cells that are ....(14) exposed to a signal often modulate their response by a ....(15) of the receptor or by reducing the number of functional receptors in the membrane in a process known as ....(16) regulation. Of the 19 hormones listed in Table 26.1, eight are specifically mentioned as involving cAMP as a ....(17) messenger. The cAMP pathway is involved in so many different effects in different cells because the response to cAMP in a given cell is specialized for what that cell is capable of recognizing via its specific ....(18). The enzyme adenylate ....(19) catalyses the transformation of ATP into 3′, 5′-cAMP.

Answers: (1) zinc; (2) Hsps; (3) chaperones; (4) dissociate; (5) nucleus; (6) element; (7) palindrome; (8) membrane; (9) activates; (10) dimerization; (11) phosphorylation; (12) dephosphorylation; (13) phosphatases; (14) repeatedly; (15) desensitization; (16) down; (17) second; (18) receptors; (19) cyclase.

Complete the following

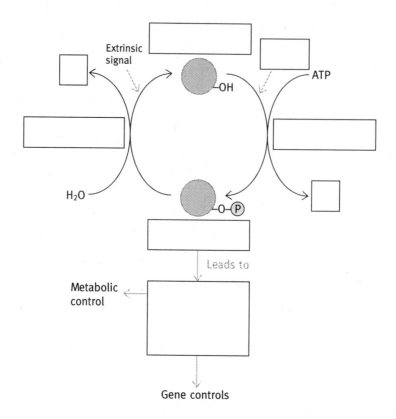

G proteins

The cytoplasmic face of the β_2-adrenergic receptor protein for epinephrine is a regulatory ....(1) protein which has three subunits α, β, γ. The α subunit can bind either ....(2) or ....(3), the latter representing the ....(4) state. When epinephrine binds to the receptor, a conformational change results in both the receptor and the G protein, which causes the GDP to be ....(5) for GTP. In this system the G protein attached to the receptor is a G_s, 's' for ....(6).   The α subunit of the G_s with the GTP molecule, ...(7),   leaves the G_s complex and activates ....(8) cyclase which produces ....(9).   The α-GTP subunit also has ....(10) enzyme activity, that is, it hydrolyses the GTP to GDP + ...(11).   The activity is low, so there is a delay before this happens to an appreciable extent. Once this reaction occurs, the α subunit contains a ....(12) molecule and it ....(13) from the adenylate cyclase (which becomes inactive) and rejoins the β and γ ....(14). If the receptor is still occupied by a hormone, the process will be repeated. There is an ....(15) effect in that more than one G protein can be activated by the ....(16) receptor complex. Cholera is responsible for ....(17) GTPase activity of the α subunit so that the signal cannot be switched off. The ....(18) adrenergic receptor for epinephrine is associated with a G_i, ...(19), protein complex which inhibits adenylate cyclase activity, the α_1-adrenergic receptor does not use cAMP but rather causes Ca^{2+} release by the ....(20) pathway. Thus this one hormone, epinephrine, can exert quite different effects according to the type of receptor. cAMP effects gene transcription through promoters containing ....(21), cAMP-responsive elements. The mechanism involves a ....(22), CRE binding protein which, when phosphorylated by protein kinase A, ...(23), becomes an active ....(24) factor of the leucine ....(25) type by complexing to CREs and promoting specific gene transcription.

Answers: (1) G; (2) GTP; (3) GDP; (4) unstimulated; (5) exchanged; (6) stimulatory; (7) α - GTP; (8) adenylate; (9) cAMP; (10) GTPase; (11) P_i; (12) GDP; (13) detaches; (14) subunits; (15) amplification; (16) hormone–; (17) inactive; (18) α_2; (19) inhibitory; (20) phosphoinositide; (21) CREs; (22) CREB; (23) PKA; (24) transcriptional; (25) zipper.

Complete the following

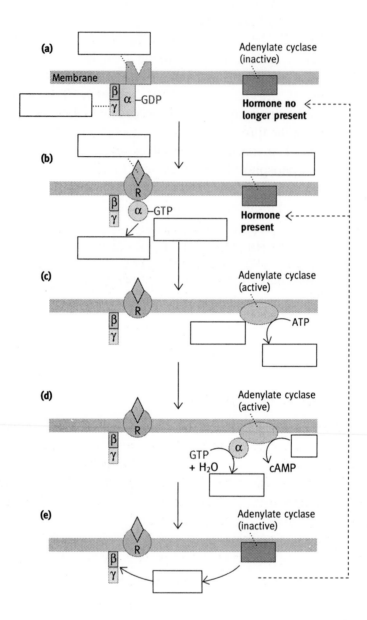

Complete the following

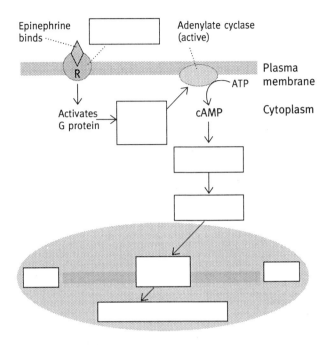

cGMP and nitric oxide (NO)

3′, 5′-cyclic GMP, ...(1), can also function as a second messenger by activating protein ....(2). In one pathway, the inner membrane domain of the receptor itself has guanylate ....(3) activity (synthesizes cGMP). A second guanylate cyclase can bind ....(4) oxide (NO) to a heme prosthetic group. The NO molecule is produced from ....(5) by the enzyme nitric oxide synthase in cells lining parts of the ....(6) system. Increased [NO] causes blood vessel ....(7). NO is oxidized to NO_2 and NO_3 in ....(8) so its effects are very localized.

Answers: (1) cGMP; (2) kinases; (3) cyclase; (4) nitric; (5) arginine; (6) vascular; (7) dilatation; (8) seconds.

Phosphatidylinositol cascade

The phosphatidylinositol cascade does not involve ....(1) nucleotides. Three hormones that operate by this mechanism are ....(2) hormone, ...(3) releasing hormone, and ....(4), platelet-derived growth factor. As in the other second messenger stories, the hormone binds to the receptor which activates a ......(5) protein, (in this case a G_p), which exchanges its

GDP for a ....(6) and activates a membrane-bound ....(7). The enzyme activated is ....(8), PLC, which catalyses the hydrolysis of ....(9), phosphatidylinositol-4,5-bisphosphate, into ....(10), inositol trisphosphate, and .....(11), diacylglycerol. The ....(12) activity of the G protein subunit makes the stimulation transitory. The IP_3 released causes ....(13) channels in the .....(14) membrane to open and ....(15) from the lumen is released into the cytosol. DAG activates protein kinase C, ...(16), which can exert multiple cellular effects via the ....(17) of target proteins. Ca^{2+} is required for ....(18) activation by DAG. Phorbol esters, which promote ....(19), are structurally similar to ....(20). They activate phosphokinase C via a ....(21) stimulation, whereas PKC stimulation via DAG is rapidly turned off due to the rapid destruction of DAG. The physiological role of Ca^{2+} as another second messenger usually involves the protein ....(22), which has four binding sites for Ca^{2+}. There are a number of calmodulin–Ca^{2+}-activated protein ....(23) that target enzymes such as glycogen ....(24) and glycogen ....(25).

Answers: (1) cyclic; (2) thyrotropin-releasing; (3) gonadotropin-; (4) PDGF; (5) G; (6) GTP; (7) enzyme; (8) phospholipase C; (9) PIP_2; (10) IP_3; (11) DAG; (12) GTPase; (13) ligand-gated; (14) ER; (15) Ca^{2+}; (16) PKC; (17) phosphorylation; (18) maximum; (19) tumours; (20) diacylglycerols; (21) prolonged; (22) calmodulin; (23) kinases; (24) phosphorylase; (25) synthetase.

Complete the following

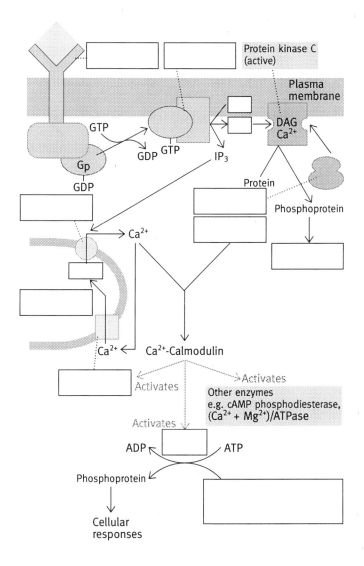

Tyrosine kinase activity

Receptors associated with ....(1) kinase activity can phosphorylate
......................(2) on the amino acid tyrosine. The PKA and PKC of other messenger pathways catalyse the addition of a phosphate from ATP to the ....(3)
groups of either serine or ....(4). Most growth factors, for example,
epidermal growth factor, and certain hormones such as insulin have ....(5)
kinase-associated receptors. When the EGF, ...(6)   growth factor,

.....................(7) is bound by its ligand (EGF), the receptors form ....(8) which results in multiple ....(9) of the tyrosine residues on these receptors by the receptors themselves. This event then activates a signal pathway that results in the activation of ....(10) factors. The Ra*s* protein is a ....(11) protein with ....(12) activity. The name Ra*s* is derived from the name of the ....(13) (*Rat Sarcoma*) because of the connection between this protein and an abnormal form called for by this ....(14). Two cytosolic proteins, ...(15), growth factor-binding protein, and SOS, son of sevenless, are also involved in the Ra*s* pathway. The GRB in association with SOS binds to the phosphorylated receptor by way of the ....(16) domain on GRB. The GTPase activity of the G proteins functions as a molecular clock to limit the activation response. The GRB–SOS-activated receptor complex causes the ....(17) protein to exchange its GDP for a ....(18),   and, during the time before this Ras hydrolyses the GTP to GDP, a protein kinase called ....(19) becomes active and starts to catalyse the ....(20) of proteins on their serine/threonine amino acids. This phosphorylation leads to a ....(21) in which transcriptional ....(22) become phosphorylated and attached to responsive elements of the gene ....(23).

Answers: (1) tyrosine; (2) themselves; (3) hydroxyl; (4) threonine; (5) tyrosine; (6) epidermal; (7) receptor; (8) dimers; (9) phosphorylation; (10) transcriptional; (11) GTP-binding; (12) GTPase; (13) virus; (14) retrovirus; (15) GRB; (16) SH2; (17) Ras; (18) GTP; (19) Raf; (20) phosphorylation; (21) cascade; (22) factors; (23) promoters.

Complete the following

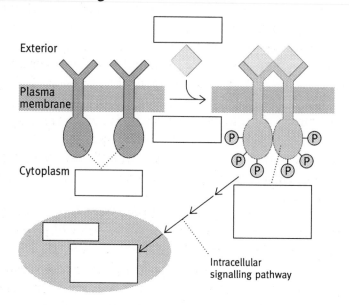

Complete the following

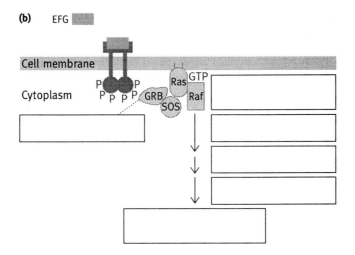

Oncogenes

The proteins Ras and Raf are ....(1) products, which means that they are normal cellular proteins that are very similar to ....(2) cancer-producing proteins that are coded for by ....(3). The oncogenes are mutated forms of the normal gene which the cell can pick up upon infection by the ....(4) or by some ....(5) event. The abnormal proteins that are coded by the oncogenes are ....(6). These abnormal oncogene product proteins for Ras, Raf, and associated pathway proteins are in a permanently ....(7) state, without the presence of signal bound to the receptor. One such oncogene, v-*erbB*, codes for an abnormal ....(8) receptor, which lacks an external domain and has a cytoplasmic domain that is permanently ....(9).   There are several mechanisms that can cause the conversion of protooncogenes to ....(10).   Many of the changes in protein sequence could have resulted in single-base ....(11) of the DNA. Chromosomal rearrangements can result in ....(12) of genes involved in the ....(13) of cell differentiation and cell division. This can cause oncogene formation in several ways: via overexpression of a protein because the gene has been placed under the control of a different ....(14) element; via oncogenic hybrid proteins caused by gene ....(15);   or via overproduction of a control protein by gene ....(16).   Cancer is the result of a cell's acquiring

.................(17) genetic abnormalities and about half of human cancers involve the ....(18) of a functional tumour ....(19) gene, specifically the ....(20) gene which, among other things, stimulates DNA ....(21) mechanisms and controls the transition between G_1 and ....(22) phase in the cell cycle. There is evidence that there are various interconnection points between the ....(23) pathways. One example is ....(24), which is activated by the ....(25) pathway and phosphorylates ....(26); another might be the activation of other proteins with ....(27) domains by the same phosphorylated receptor dimer.

Answers: (1) protooncogene; (2) abnormal; (3) retroviruses; (4) virus; (5) mutation; (6) dysfunctional; (7) activated; (8) EGF; (9) phosphorylated; (10) oncogenes; (11) mutations; (12) translocation; (13) control; (14) regulatory; (15) fusion; (16) amplification; (17) multiple; (18) loss; (19) suppressor; (20) p53; (21) repair; (22) S; (23) different; (24) PKC; (25) PIP_2; (26) Raf; (27) SH2.

Interferons

Interferons are involved in a direct ....(1) signalling pathway from the receptor to the ....(2). Interferons were first recognized for their ....(3) activity, in which the cell activates certain genes through ....(4), interferon-stimulated response elements. The receptor for γ-interferon has two cytoplasmic tyrosine ....(5) molecules, called ....(6) kinases, associated with it. These ....(7) and become ....(8) upon receptor stimulation. Two proteins from the cytosol, which also have the same ....(9) domains, bind to the phosphorylated dimer and become phosphorylated; they then move into the ....(10), assemble into an active ....(11) factor and promote the transcription of ....(12) elements.

Answers: (1) intracellular; (2) nucleus; (3) antiviral; (4) ISREs; (5) kinase; (6) JAK; (7) dimerize; (8) phosphorylated; (9) SH2; (10) nucleus; (11) transcriptional; (12) ISRE.

Complete the following

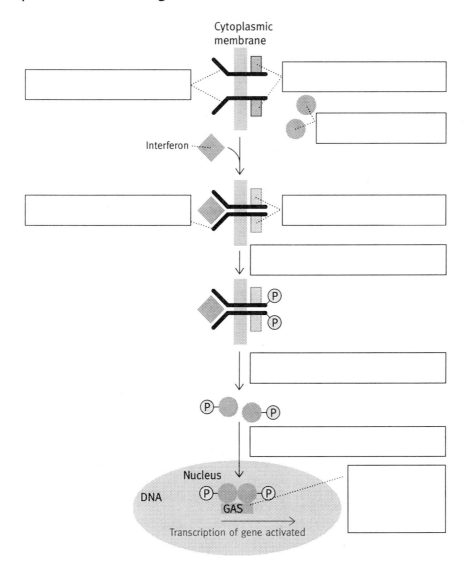

Ligand-gated ion channels

Ion channels in membranes can be controlled by ....(1) binding to a receptor. One such example is via ....(2) signalling as found in the contraction of ....(3)   muscle in response to ....(4)   released from the ....(5) neurone. Neurones have relatively ....(6)   K$^+$ levels and low ......(7) levels ....(8)   as compared to outside the cell, which is the result of the action of the Na$^+$/K$^+$ ATPase ion pump. Due to differences in the

number of ....(9) particles inside the cell versus the number outside (some
K^+ leaks out), the membrane is ....(10) with a (+) outside charge and a (–)
inside charge resulting in a resting cell potential of about ....(11)
When ....(12) binds to its receptor, the receptor ....(13)
as a channel and allows ......(14) and K^+ to move through. These ions naturally
move 'downhill' in the direction of the concentration gradient: Na^+ moves ......(15);
K^+ moves .....(16). The Na^+ moves in at a faster rate than the K^+ moves out, and the
membrane loses some of the charge built up and becomes ....(17).

Answers: (1) ligand; (2) neuronal; (3) striated; (4) acetylcholine; (5) presynaptic; (6) high; (7)
Na^+; (8) inside; (9) charged; (10) polarized; (11) –70 millivolts; (12) acetylcholine; (13) opens; (14)
Na^+; (15) in; (16) out; (17) depolarized.

Complete the following

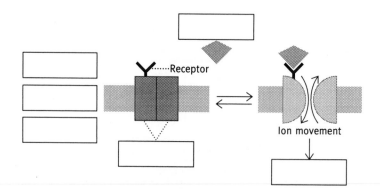

Voltage-gated ion channels

The ....(1) impulse ....(2) along the ....(3) by a process in which
different types of Na^+ and K^+ channels, called ....(4) channels,
open and close in response to changes in the local membrane potential, thus generating a ....(5) of membrane depolarization from the ....(6) to the
.................(7) endings. At the ends of the neurones are voltage-gated ....(8)
channels which permit extracellular Ca^{2+} to enter in response to this wave of
.................................(9). This change in the Ca^{2+} concentration results in
acetylcholine release by ....(10). Muscle cell receptors for acetylcholine respond, and the initial Ca^{2+} gradient is restored with a ....(11) ATPase.
This Ca^{2+}-stimulated exocytosis of vesicle contents has been found to occur in many
other biological mechanisms. The process of vision depends on the control of
.......................(12) channels. There are two types of cells for light detection in
the retina of vertebrates, ...(13) and ....(14); each has a specific role
in vision. In the dark, rod cells have ....(15) cGMP which keeps ligand-gated
cation channels ....(16). The cell potential is about –30 mV. Rhodopsin in the

rod cells is a complex of opsin and 11-*cis*-retinol. Light changes 11-*cis*-retinol to all-
...............(17) retinol which results in a significant conformational change. Look
closely at where carbon 13 is with respect to carbon 12 in the 11-*cis* and all-*trans*
structures of retinol. Through a G protein activation sequence the conformational
change in rhodopsin causes the enzyme cGMP ....(18) to
hydrolyse cGMP which causes the Na⁺ channel to ....(19); an
...................(20) in the membrane polarization then occurs, known as
hyperpolarization. The recovery of the cell after illumination involves inactivation of
...................(21), lowering of cellular Ca^{2+} levels, ...(22) cGMP
levels, and recycling of the rhodopsin. The Na⁺ channel is ....(23), and
the membrane potential returns to resting state.

Answers: (1) nerve; (2) moves; (3) axon; (4) voltage-gated; (5) wave; (6) axon; (7) neuronal;
(8) Ca^{2+}; (9) depolarization; (10) exocytosis; (11) Ca^{2+}-; (12) ligand-gated; (13) rods; (14) cones;
(15) high; (16) open; (17) *trans*-; (18) phosphodiesterase; (19) close; (20) increase; (21) rhodopsin;
(22) increased; (23) opened.

Complete the following

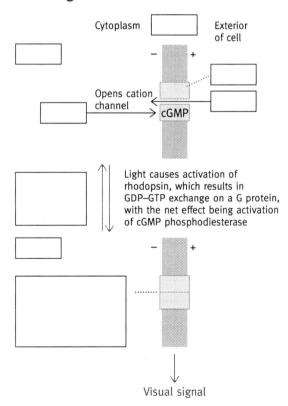

Review of problems from the end of Chapter 26

- There are three classes of molecules responsible for carrying signals between cells: hormones; growth factors; and neurotransmitters.

- Lipid-soluble molecules are nonpolar and as such are not water-soluble. These molecules are able to cross the cell membrane without involving a membrane receptor. These molecules do interact with an intracellular receptor before conveying their signal to the nucleus. Lipid-insoluble molecules are polar and water-soluble and do not cross the cell membrane. The signal is carried into the cell via a second messenger by a mechanism that involves a membrane receptor.

- Cyclic AMP (cAMP) is produced by the action of the enzyme adeylate cyclase on ATP. This enzyme becomes active when an α subunit from the trimeric G protein complexed with a GTP molecule binds with the inactive adenylate cyclase. It is the action of the GTP displacing the GDP from the α subunit while it is a part of the trimeric (α, β, γ) complex that causes the α subunit to separate from the β and γ subunits and bind with adenylate cyclase. The GTP is hydrolysed to GDP and P_i by the α subunit itself. This event causes the α-GDP complex to reform the trimeric α, β, γ complex and cause the adenylate cyclase to again be inactive. The action of cholera is to block this GTPase activity so that the adenylate cyclase stays activated and the levels of cAMP stay elevated.

- cAMP is an allosteric activator of protein kinase A that phosphorylates enzyme proteins, as well as transcription factors; the latter are cAMP-responsive elements (CREs), which are parts of specific genes.

- Nitric oxide, NO, activates guanylate cyclase which produces the second messenger cGMP.

- IP_3 and DAG are second messengers released through the phosphatidylinositol cascade pathway.

- The Ras pathway does not involve a small-molecular-weight second messenger; the Raf protein kinase is thought to be activated directly by the membrane-bound Ras which is activated by the integral membrane epidermal growth factor (EGF) receptor.

- Proteins with SH2 domains associate with different autophosphorylated sites on dimerized receptors and function to link multiple pathways to an activated (phosphorylated) receptor.

- Signals that are either always on or always off are like switches that are stuck in one position. Since these signals often control some aspect of cell division, the connection with cancer comes from uncontrolled cell division.

- Interferon can activate specific gene transcription through a receptor that is phosphorylated by a protein dimer known as the Janus kinases (JAK), and then phosphorylates STAT (signal transducer and activator of transcription) proteins which activate the γ-interferon-activated sequence element (GAS) on the gene that is affected.

• Make a table to help keep receptor type and mechanism of activation ideas organized.

Type of receptor	Mechanism of activation
Epinephrine receptor activating adenylate cyclase	The receptor changes conformation and the α subunit of a G protein binds ATP and interacts with the membrane-bound adenylate cyclase
EGF receptor of the Ras pathway	The receptor dimerizes; this activates a self-phosphorylation of tyrosine residues, by an intrinsic kinase
Interferon receptor	The receptor dimerizes and a separate tyrosine kinase phosphorylates the receptor. The phosphorylated receptor is now active as a kinase and phosphorylates STAT proteins which bind to the GAS promoter

• A voltage-gated Ca^{2+} channel opens when a wave of neuronal membrane depolarization reaches it. An example is the nerve synapse that liberates acetylcholine.

• A ligand-gated channel opens when a specific ligand binds to it. The cation channels of the rod cell membranes are an example.

Additional questions for Chapter 26

1. Complete the following table of abbreviations. (*Hint.* Use the list of abbreviations found at the beginning of the text.)

AMP	?
?	Cyclic AMP
cGMP	?
CSF	?
TSH	?
ACTH	?
?	Luteinizing hormone
?	Follicle-stimulating hormone
?	Antidiuretic hormone
?	Thyroxine
?	Triiodothyronine

?	Heat shock proteins
G	?
G_s	?
α-GTP	?
PKA	?
CREs	?
CREB	?
NO	?
AC	?
?	Plasma membrane guanylate cyclase
?	Cytoplasmic guanylate cyclase
?	Phosphatidylinositol-4,5-bisphosphate
?	G protein involved in PIP_2 pathway
?	Phospholipase C
?	Inositol triphosphate
?	Diacylglycerol
ER	?
Ras	?
GRB	?
SOS	?
SH2	?
Src	?
Raf	?
?	Epidermal growth factor
?	GTPase activating proteins
?	Oncogene that codes for abnormal EGF receptor
?	Gene involved in Burkitt's lymphoma
?	Tumour suppressor gene
?	Interferon-stimulated response element
?	Janus kinase type of tyrosine kinase
GAS	?
STAT	?

2. Why is it that lipid-soluble signalling ligands do not require a membrane receptor?

3. List at least eight hormones that cause a cellular response through a cAMP-mediated mechanism.

4. What is the difference between paracine and autocrine action?

5. What are the two structural differences between epinephrine and norepinephrine?

6. What is the signal connection between the hypothalamus and the thyroid?

7. Explain the rationale behind the statement, 'The essence of control is reversibility.'

8. What common membrane component has a structure similar to those of testosterone and progesterone? What are the major structural differences and how do these differences affect these molecules' properties?

9. What is the role of the heat shock proteins. Hsps, in the mechanism of action of the steroid/thyroxine receptors?

10. What two alternative events are common in response to a signal ligand binding to a membrane receptor?

11. What are two mechanisms whereby an extrinsic signal can cancel out the effects of a second messenger of a pathway that has already been activated?

12. How many transmembrane loops are there in the β_2-adrenergic receptor?

13. What would happen if the α-GTP subunit GTPase activity occurred at a faster rate than that at which this subunit could find an adenylate cyclase?

14. What is needed for the CREB to interact with the CRE and how does this change come about?

15. In general, how do the effects of cGMP compare with those of cAMP? Give an example.

16. How is NO produced?

17. What are two rather unique chemical structural elements of phorbol esters?

18. How many Ca^{2+} binding sites are there on calmodulin?

19. What three second messengers are associated with the phosphatidylinositol cascade?

20. What events occur in the process of protein kinase C becoming active?

21. Where does the Ca^{2+} come from when released in response to the opening of the IP_3-gated channel? How does it get back?

22. What is the SH2 domain and what protein of the Ras pathway has this?

23. What does the abbreviation MAPK stand for and where do MAPKs fit into the Ras pathway?

24. How is the Ras activation of Raf protein kinases turned off and what is the role of GAPs?

25. What would be one way to link multiple response pathways to an activated phosphorylated dimeric receptor?

26. What is the role of the STAT protein in the γ-interferon signalling pathway?

27. What are the normal relative amounts of Na^+ and K^+ between the inside and outside of most animal cells?

28. What is the role of Ca^{2+} in the process of acetylcholine stimulation of a motor neurone?

29. What is the relative level of cGMP in rod cells in the dark? What is the consequence of this situation?

30. What is the role of the cGMP phosphodiesterase in the process of vision?

Chapter 27

The red blood cell and the role of hemoglobin

Chapter summary

This chapter describes the chemical structure of the heme molecule, the importance of the hemoglobin protein, and the role of the red blood cell in the transport of oxygen and CO_2. The enzymes involved in heme biosynthesis and the role of Fe levels in controlling this process are highlighted. The binding of oxygen by hemoglobin is described in detail and compared with that of myoglobin. The natures of the tense and relaxed states are explained, as well as the role of 2:3-bisphosphoglycerate (BPG). The chemical reactions associated with the Bohr effect and the chloride shift and their relationship to the transport of CO_2 and the buffering of blood pH are outlined. The chapter ends with a description of the biochemistry behind the occurrence of sickle cell anemia.

Learning objectives

- ❏ The overall importance of the hemoglobin protein and the red blood cell in delivering oxygen from the lungs to the tissues.

- ❏ The chemical nature of the heme molecule including the protoporphyrin structure that contains four pyrrole rings.

- ❏ An overview of the synthesis of heme.

- ❏ The nature of the binding that holds the Fe^{2+} in position in the heme.

- ❏ The mechanism by which the levels of Fe affect the synthesis of heme via an IRE.

- ❏ The roles of transferrin and the levels of its receptor in the mechanism and control of Fe transport

- ❏ The mechanism by which the levels of Fe influence the synthesis of the transferrin protein receptor.

- ❏ The mode in which iron is stored in red blood cells.

❑ The average lifetime of red blood cells and how to distinguish between old cells and young cells.

❑ The origin of bilirubin and its ultimate fate.

❑ The control mechanism associated with the production of the globin part of hemoglobin and the role of eIF2.

❑ The importance of the oxidation state of Fe in relationship to its ability to bind oxygen.

❑ The role of myoglobin as a storage protein for oxygen.

❑ The similarities and differences between hemoglobin and myoglobin, especially in their ability to bind oxygen

❑ The difference between a sigmoid saturation curve and a hyperbolic one, and a possible explanation for the occurrence of the latter.

❑ The physiological significance of the change in the affinity of hemoglobin for oxygen as a function of the O_2 concentration.

❑ The natures of the tense and relaxed conformational states for hemoglobin subunits and their relative affinities for O_2, and the effect of O_2 binding on their relative stabilities.

❑ The effect of BPG on the oxygen binding affinity of hemoglobin and the physiological significance of this molecule.

❑ The chemical reaction associated with the Bohr effect and the connection between this reaction and the transport of CO_2 by red blood cells.

❑ The process known as the chloride shift and its relationship to CO_2 transport.

❑ The importance of the bicarbonate ion in buffering blood pH.

❑ The biochemical explanation for the occurrence of sickle cell anemia, and the effect of this disease on the binding of O_2 by red blood cells.

A walk through the chapter

Red blood cells

Hemoglobin is important in the process of delivering ....(1) from the lungs to the tissues of the body. About 160 million red cells, ...(2), are produced every ....(3) and circulate for about .....(4) days. The characteristic shape of the cell with its large surface area facilitates the exchange of ....(5) in and out. Red cells come from hemopoietic stem cells via a process stimulated by ....(6), the concentration of which is ....(7)

responsive to the O_2 levels. Immature red cells are called ....(8); they undergo many changes during the maturation process.

Answers: (1) oxygen; (2) erythrocytes; (3) minute; (4) 100; (5) gases; (6) erythropoietin; (7) inversely; (8) reticulocytes.

Heme synthesis

The heme is a relatively small organic molecule that serves as the site of ....(1) attachment. The protoporphyrin structure, a tetrapyrrole, contains four ....(2) rings, each being composed of four carbons and one nitrogen atom. The pyrroles are ....(3) with extra cyclic attachments and are connected together by one-carbon bridges. The double bonds of the protoporphyrin are ....(4). The iron forms bonds with the four ....(5) of the pyrroles. The iron of the heme is in the ....(6) state and is called ferrous; Fe^{3+} is called ferric. The heme is synthesized from ....(7) and ....(8) CoA, which are joined by the enzyme ALA-S, ...(9) synthase, to produce 5-aminolevulinate. In the steps leading to heme, two of these product molecules are dehydrated to form .....(10), porphobilinogen, via the enzyme ....(11). Four of these PBG molecules are linked, the side groups are modified, and an Fe^{2+} atom is inserted and chelated, with the first and last step occurring in the ....(12). The rate-limiting step of heme synthesis is the ALA ....(13) activity, and Fe levels affect the expression of this enzyme by an IRE, ...(14) element, on the mRNA, which interacts with an IRE binding ....(15) at low iron levels, and allows translation of the protein. The expression of the ALA synthase protein is ....(16) under conditions of ....(17) levels of iron, because the IRE binding protein does not interact with the ....(18) on the mRNA and ....(19) is stopped.

Answers: (1) iron; (2) pyrrole; (3) substituted; (4) conjugated; (5) nitrogens; (6) Fe^{2+}; (7) glycine; (8) succinyl-; (9) aminolevulinate; (10) PBG; (11) ALA-dehydratase; (12) mitochondria; (13) synthase; (14) iron-responsive; (15) protein; (16) blocked; (17) high; (18) IRE; (19) translation.

Iron transport and use

Transferrin is a blood plasma protein that transports ....(1) and is taken into the cell by receptor-mediated ....(2). Cellular iron ....(3) are controlled by the amount of transferrin ....(4) present on the outside of the cell. The ....(5) of the receptor protein is controlled through the level of cellular ....(6) by the regulation of the ....(7) of the ....(8) for the receptor. This stability comes from the ....(9) protein, which binds to the mRNA in the ....(10) of iron and protects the mRNA from ....(11). As the levels of iron ....(12), the IRE-binding protein no longer protects the mRNA which becomes ....(13) so that less receptor is available to mediate the uptake of iron–transferrin. This process is generally known as ....(14) control. Iron is stored in red cells as a complex of ....(15) and inorganic iron. The amount of the protein apoferritin is regulated by iron levels such that iron causes a ....(16) of an IRE-binding protein from the mRNA which then ....(17) the synthesis of the protein. After about 110 days, the red cell is aged and ready for destruction. Old cells are distinguished from young cells by the ....(18) of the membrane ....(19). The red cells are destroyed mainly by ....(20) cells, and the iron is ....(21) through a process in which the ring is opened by the enzyme heme ....(22). The organic portion of the heme (without Fe) is converted to ....(23), which is transported by serum albumin, ...(24) to become more polar, and excreted, becoming part of the yellow-coloured component of urine.

Answers: (1) iron; (2) endocytosis; (3) levels; (4) receptor; (5) synthesis; (6) iron; (7) stability; (8) mRNA; (9) IRE-binding; (10) absence; (11) degradation; (12) increase; (13) degraded; (14) feedback; (15) apoferritin; (16) detachment; (17) permits; (18) composition; (19) glycoproteins; (20) reticuloendothelial; (21) recycled; (22) oxygenase; (23) bilirubin; (24) glucuronated.

Myoglobin

The globulin part of hemoglobin is a ....(1), which is synthesized only when ....(2) is available. The mechanism for this control involves a protein ....(3), which in the ....(4) of heme phosphorylates ....(5), which is a protein factor needed for the general ....(6) of mRNA translation in eukaryotes. The phosphorylated eIF2 cannot play its role in the initiation of protein synthesis. Ferrous (Fe^{2+}) but not ferric (Fe^{3+}) ions can combine in the needed way with ....(7). The Fe^{3+} form of heme is called ....(8). Iron in the heme in the crevice of the hemoglobin protein molecule is ....(9) in the Fe^{2+} state. In addition to the ....(10) ligand sites occupied by the ....(11), oxygen binds to one of the empty ligand-binding sites, and a ....(12) residue occupies the final one. Myoglobin is a ....(13) protein involved in oxygen ....(14) storage. It is similar to hemoglobin, except that it consists of a ....(15) heme bound to a single protein molecule. The graph of the ....(16) percentage saturation that is attained by the protein myoglobin as a function of the pressure of oxygen, ...(17), is very important. The pO_2 can be thought of as a measure of concentration. On the same curve is shown the response of hemoglobin. Notice that at ....(18) values of pO_2, the % saturation of myoglobin is ....(19) than that of hemoglobin. This means that myoglobin has a

......................(20) affinity for oxygen than does hemoglobin at these levels of O_2, and so would readily take the ....(21) from ....(22), and make it available in the ....(23) cell.

Answers: (1) protein; (2) heme; (3) kinase; (4) absence; (5) eIF2; (6) initiation; (7) oxygen; (8) hematin; (9) stable; (10) four; (11) pyrroles; (12) histidine; (13) muscle; (14) reserve; (15) single; (16) relative; (17) pO_2; (18) low; (19) greater; (20) higher; (21) O_2; (22) hemoglobin; (23) muscle.

Oxygen binding in hemoglobin

Hemoglobin has ....(1) subunits, each of which resembles the ....(2) monomer, with each having a heme capable of binding oxygen. The subunits are not all identical, but rather there are two α and two β subunits. A molecule of hemoglobin can bind four .....(3) molecules. The oxygen binding curve for hemoglobin is a ....(4) curve and shifted to the ....(5), higher pO_2, with respect to that for myoglobin. The affinity of hemoglobin for oxygen is ....(6) than that for myoblogin, which means that the oxygen concentrations required to ....(7) saturate hemoglobin are higher than those required for the same level of ....(8) for myoglobin. The response curve is appropriate for hemoglobin, which must release its ....(9) most efficiently at the pressure of oxygen encountered in the ....(10). The pO_2 in the ....(11) is very high so, as the hemoglobin is circulated back there, it will pick up a full supply. The ....(12) oxygen saturation curve is achieved because hemoglobin is a ....(13) protein that undergoes a ....(14) change upon the binding of O_2. This change in structure results in a higher ....(15) of the other subunits for additional O_2 molecules. This is known as a homotropic positive ....(16) effect. The cooperative binding effect in hemoglobin can be explained in terms of the principles of the ....(17) model, although a clear distinction between the concerted and ....(18) models cannot be made in this case. There are two conformational states for hemoglobin: the ....(19), T, state with ....(20) O_2 affinity and the ....(21), R, state with ....(22) oxygen affinity. At low O_2 concentrations the ......(23) state predominates but, as the level of O_2 increases and the sites become occupied, there is a swing toward the .....(24) conformation. One driving force for this conformational change is the fact that, on O_2 binding to the ....(25), the iron atom moves more into the..........(26) of the tetrapyrrole ring. The polypeptide is attached directly to the iron through a ....(27) residue.

The small, doubly phosphorylated, three-carbon molecule BPG, 2:3-bisphosphoglycerate, plays an important role in oxygen transport. This molecule can cause a shift in the O_2 binding curve to the ....(28) corresponding to the higher pO_2 needed to saturate the molecule; thus a ....(29) affinity of hemoglobin for O_2. The effect of lowering the affinity of hemoglobin for O_2 means that there is an

....................(30) amount of O_2 unloaded to the tissues. The binding of the negatively charged BPG to a ....(31) charged cavity in the hemoglobin causes the protein to stay in the unoxygenated T state. The oxygenated R state does not interact with BPG. The levels of BPG are regulated in accord with oxygen levels in the tissue. BPG also plays a part in the hemoglobin of a developing ....(32). The fetal hemoglobin subunit is different from that of the adult and interacts less strongly with BPG, resulting in a ....(33) affinity for O_2 than the adult hemoglobin at the same levels of BPG; this acts as a driving force for the transfer of O_2 from the mother to the fetus.

Answers: (1) four; (2) myoglobin; (3) O_2; (4) sigmoidal; (5) right; (6) lower; (7) 50%; (8) saturation; (9) oxygen; (10) capillaries; (11) lung; (12) sigmoidal; (13) multisubunit; (14) conformational; (15) affinity; (16) cooperative; (17) concerted; (18) sequential; (19) tense; (20) low; (21) relaxed; (22) high; (23) T; (24) R; (25) heme; (26) plane; (27) histidine; (28) right; (29) lower; (30) increased; (31) positively; (32) fetus; (33) higher.

Complete the following

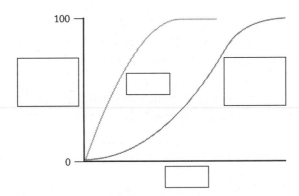

The Bohr effect

When O_2 binds to hemoglobin protons are ....(1) in a phenomenon known as the Bohr effect. This effect can also be described by saying that the R, oxygenated form is a stronger ....(2) than the T, deoxygenated form. Carbon dioxide, ...(3), is involved in the Bohr effect because the hydrated form of carbon dioxide is ....(4) acid, H_2CO_3, which readily loses a ....(5) at physiological pH. Carbon dioxide is produced in the tissue and enters the red blood where the enzyme carbonic ....(6) converts it to H_2CO_3. By way of the Bohr effect, the ......(7) from H_2CO_3 causes the hemoglobin molecule to ....(8) more O_2. In the red cell the HCO_3^-, ...(9) anion, exits via an ion ....(10) and in response a

.......(11) enters the cell. This process is known as the chloride ....(12). The Bohr effect of proton release accompanying the binding of ....(13) by hemoglobin is also involved at the ....(14), where the H^+ released reprotonates the bicarbonate ion, ...(15), to form carbonic acid, which ....(16) anhydrase can dehydrate to produce CO_2 to be ....(17). A reverse chloride effect is experienced at the tissues. Thus, the red blood cell operates as a ....(18) transport in one direction and an ....(19) transport in the other. The acid/base equilibrium between carbonic acid (H_2CO_3) and the bicarbonate ion (HCO_3^-) also helps to ....(20) the pH of the blood.

Answers: (1) released; (2) acid; (3) CO_2; (4) carbonic; (5) proton; (6) anhydrase; (7) H^+; (8) release; (9) bicarbonate; (10) channel; (11) Cl^-; (12) shift; (13) O_2; (14) lung; (15) HCO_3^-; (16) carbonic; (17) expelled; (18) CO_2; (19) O_2; (20) buffer.

Complete the following

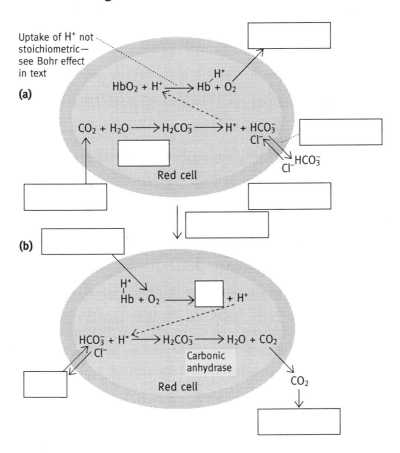

Sickle cell anemia

Sickle cell ....(1) is a classic biochemical example of how a single amino acid change in the sequence of a protein can have a profound effect. In normal hemoglobin, amino acid 6 of the ....(2) is glutamic acid which is ....(3) charged. The codons in mRNA for glutamic acid are GAA and GAG. In sickle cell anemia, a mutation in the central base of either of these codons from A to U charges the amino acid to ....(4). Since valine is a ....(5) amino acid and thus very different from glutamic acid, this substitution changes the way in which the protein chains interact. The abnormal hydrophobic patch on the sickle cell hemoglobin caused by valine allows the ....(6) hemoglobin molecules to bind to each other resulting in a long multistrand rigid rod that distorts the normal ....(7) shape of the cell into a sickle shape. This structure tends to block the ....(8). Interestingly, there is some evidence that this mutation offers protection against ....(9).

Answers: (1) anemia; (2) β-chain; (3) negatively; (4) valine; (5) nonpolar; (6) deoxygenated; (7) disc; (8) capillaries; (9) malaria.

Review of problems from the end of Chapter 27

- Heme biosynthesis involves 5-aminolevulinate (ALA) synthase and ALA dehydratase. The first step in heme biosynthesis involves coupling the citric acid cycle metabolite succinyl-CoA with the amino acid glycine to produce 5-aminolevulinic acid (ALA) in a reaction catalysed by ALA synthase. In the second reaction two molecules of ALA are joined with the loss of two molecules of water to form porphobilinogen in a reaction catalysed by the enzyme ALA dehydratase.

- Increased production of ALA is associated with acute intermittent porphyria also known as 'mad king' disease.

- The mRNA coding for ALA synthase has an iron-responsive element (IRE) which can interact with an IRE-binding protein when the concentration of iron in the cell is low. The complex formed between the IRE and its binding protein blocks the translation of the mRNA so that no ALA synthase protein is produced.

- The myoglobin protein becomes saturated with O_2 at a lower pO_2 than the hemoglobin which reflects the fact that myoglobin has a higher affinity for O_2. This is apparent by the fact that the myoglobin O_2 binding curve is further to the left. The qualitative shapes of the two curves are also different, with the myoglobin exhibiting the more expected hyperbolic shape and the hemoglobin exhibiting a more unique S-shaped sigmoidal curve.

- This sigmoidal oxygen-binding curve seen with hemoglobin is known as cooperativity and is similar in principle to the process seen in enzymes. The hemoglobin molecule can exist in two states differing in their affinity for oxygen. At low oxygen levels the binding site is in the low-affinity state; as oxygen is bound, the system switches in proportion to the higher O_2 affinity state.

- The actual conformational changes in the protein from the T state (low-affinity) to the R state (high-affinity) are linked to changes in geometry of the Fe–porphyrin structure upon O_2 binding. Specifically, the binding of O_2 causes the iron to move further into the plane of the porphyrin ring and brings an attached histidine along with it.

- At a given concentration of 2:3-bisphosphoglycerate (BPG) the fetal form of hemoglobin has a higher affinity for oxygen than does the maternal form. The rationale for this comes from the fact that the effect of BPG in general is to lower the affinity of hemoglobin for oxygen and, while it does this to both the maternal and fetal forms, the effect is much less on the fetal form because this protein has a weakening binding interaction with BPG. Thus, a higher affinity for O_2 in the fetal hemoglobin means that the O_2 will be transferred from the maternal form in this direction.

- Red blood cells aid in the transport of CO_2 from the tissues to the lungs, as well as in the transport of O_2 in the opposite direction. This process involves a phenomenon known as the chloride shift which refers to the movement of Cl^- and HCO_3^- in opposite directions across the red blood cell plasma membrane. This dual movement balances the electrical change created in the interconversion of CO_2 and H_2CO_3 and the proton exchange which occurs in the Bohr effect involving the reaction between O_2 and hemoglobin. Notice that the reaction sequence goes in one direction at the tissue side and the other at the lung side of the transport pathway.

Additional questions for Chapter 27

1. About how many red blood cells are produced per minute in the human body?
2. What is the role of erythropoietin?
3. What are the two starting materials for heme biosynthesis?
4. How many pyrrole rings are there in a heme molecule?
5. How is iron stored and what components make up this complex?
6. What is the effect on heme synthesis when the IRE-binding protein is attached to the IRE?
7. What are the main steps involved in the conversion of heme into bilirubin?
8. Which state of the hemoglobin subunit (T or R) has the higher affinity for O_2?

9. What is the effect of BPG on the affinity of hemoglobin for oxygen?

10. What chemical reaction is associated with the Bohr effect?

11. What reaction is catalysed by the enzyme carbonic anhydrase?

12. What is the connection between the release of O_2 by hemoglobin to the tissue, the Bohr effect, and the uptake of CO_2?

13. What is the genetic abnormality associated with sickle cell anemia?

Chapter 28

....

Muscle contraction

Chapter summary

This chapter describes the mechanism of muscle contraction on the molecular level. The two main classes of muscle, smooth and striated, as well as the slow and fast twitching types of striated muscles are compared in terms of location and energy supply. The geometry of the thick and thin filaments and the roles of actin, myosin, and the hydrolysis of ATP in the process of muscle contraction are highlighted. The role of Ca^{2+} in controlling the contraction of both smooth and striated muscles is described.

Learning objectives

- ❑ The two main classes of proteins involved in muscle contraction, where they are found, how they are controlled, and their distinguishing characteristics.

- ❑ The differences between slow and fast twitching muscles in terms of location and energy usage.

- ❑ The role of creatine phosphate as a reserve source of phosphoryl groups.

- ❑ The terms applied to the various structures within striated muscles, including sarcolemma, sarcoplasmic reticulum, and sarcomeres

- ❑ The geometry of the arrangement of thick and thin filaments between the Z discs.

- ❑ The occurrence of actin in the thin filaments and myosin in the thick filaments.

- ❑ The importance of the enzyme on the head of the myosin that hydrolyses ATP to ADP and P_i.

- ❑ The sequence of events that occurs in the actual movement of the muscle fibres.

- ❑ The role of Ca^{2+} in skeletal muscle contraction.

- ❑ The role of the troponin–tropomyosin complex in controlling muscle movement.

- ❑ The differences between the control mechanism for smooth muscle and that for striated muscles.

A walk through the chapter

Classes of muscles

Type	Location (examples)	Control	Characteristics
...............(1)	Intestine and ....(2) vessels	...................(3)	Contracts ....(4) and maintains contraction for extended periods
...............(5)	Skeletal muscle	....................(6)	Contracts ....(7)

The classification of striated muscles can be further divided into fast twitching and slow twitching fibres.

Subtype	Location (example)	Characteristics
..................(8) twitching	White muscle of fish and ocular muscles	ATP produced from glycolysis of glycogen reserves; ....(9) blood supply; rapid response; and ....(10) exhausted
................(11) twitching	Human back muscles	ATP from oxidative phosphorylation; ....(12) supplied with blood; ....(13) response

Answers: (1) Smooth; (2) blood; (3) Involuntary; (4) slowly; (5) Striated; (6) Voluntary; (7) rapidly; (8) Fast; (9) poor; (10) quickly; (11) Slow; (12) richly; (13) longer.

Actin and myosin

The striated cells of skeletal muscle are called ....(1) and may be very long and contain ....(2) nuclei. The cell membrane, called the ....(3), has nerve endings associated with it. There are also many long ....(4), each surrounded by a sac called the ....(5) reticulum. The myofibril structure is divided into segments called ....(6), bounded by Z discs. The muscle contracts when the Z discs are pulled closer together. The ....(7) globular protein, G actin, is one of the main protein components of the thin filament rods called ....(8). The long polymer fibres of the ....(9) have a head-to-tail polarity. There are also thick filaments that fit inside a cage array of thin filaments. The protein of the thick filaments is ....(10), which has a coiled coil configuration of two polypeptides. Two α helices coil around each other, with two ....(11) myosin chains occurring in each head. A thick filament consists of thousands of myosin molecules arranged in a ....(12) fashion. Titin is one of the proteins involved in ....(13) the thick filaments with respect to the thin ones. The myosin head is an enzyme that hydrolyses ATP to ....(14) and P_i. The energy released by this hydrolysis reaction drives the mechanical work of muscle ....(15). However, it is after the actual hydrolysis occurs, when the product ADP molecule ....(16) the protein, that the free energy is liberated. In a sense the free energy of ATP hydrolysis is captured by the ADP-bound ....(17), and this energy results in ....(18) work as a result of the dissociation of the ADP. The myosin head is ready to exert a force when it is in the ....(19) configuration; after this exertion of force it reverts to the resting ....(20). The head attaches to the ....(21) filament in the primed state and performs a power stroke on the actin as it moves to the resting state.

Answers: (1) myofibres; (2) multiple; (3) sarcolemma; (4) myofibrils; (5) sarcoplasmic; (6) sarcomeres; (7) monomeric; (8) F actin; (9) G actin; (10) myosin; (11) light; (12) bipolar; (13) positioning; (14) ADP; (15) contraction; (16) leaves; (17) protein; (18) mechanical; (19) primed; (20) configuration; (21) actin.

Complete the following

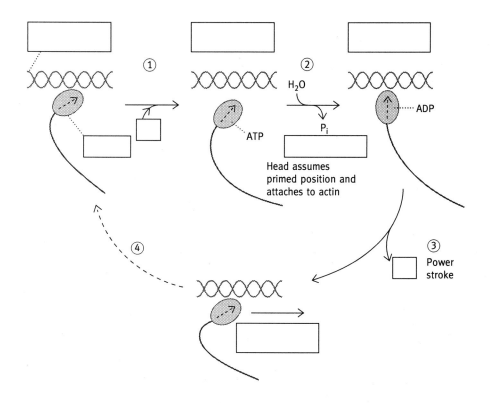

Control of skeletal muscles

In skeletal muscles, contraction is initiated by a ....(1) impulse causing ....(2) ions to move from the sarcoplasmic ....(3) to interact with the ....(4). The signal is ....(5) as the Ca^{2+} is quickly returned to the sarcoplasmic reticulum. Tropomyosin is another protein associated with ....(6) filaments. This protein has seven ....(7) attachment sites. There is a complex of three proteins called ....(8) attached to one end of the ....(9). The presence of Ca^{2+} causes this troponin–tropomyosin complex to move, ...(10) the myosin head to ....(11) to the actin filament and begin the ....(12) cycle using ATP. If the Ca^{2+} is removed, the troponin–tropomyosin complex moves ....(13) and shuts down the contraction sequence. There are two Ca^{2+} channels in the membrane of the sarcoplasmic reticulum. One is a ....(14) channel that pumps Ca^{2+} ....(15); the other is a ....(16) Ca^{2+} channel that ....(17) upon receipt of a nerve impulse and lets the Ca^{2+} flow ....(18). Acetylcholine is involved at the neuromuscular junction which causes a membrane ....(19). It is important that the electrical signal reaches all the necessary muscles in a ....(20), coordinated fashion.

Answers: (1) nerve; (2) Ca^{2+}; (3) reticulum; (4) myofibril; (5) transient; (6) thin; (7) actin; (8) troponin; (9) tropomyosin; (10) permitting; (11) attach; (12) contraction; (13) back; (14) Ca^{2+}/ATPase; (15) in; (16) voltage-gated; (17) opens; (18) out; (19) depolarization; (20) simultaneous.

Complete the following

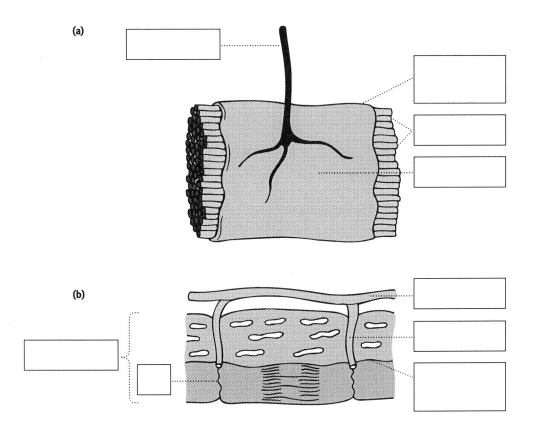

Smooth muscles

Smooth muscles form patterns that are appropriate to their ....(1), and the basic principles of contraction for these are the ....(2) as for the skeletal muscles but, instead of the sacromere structure, actin ....(3) run the length of the cell and are ....(4) at one end into the cell membrane. The control mechanism for smooth muscle is much ....(5) from that for striated muscle. A smooth muscle contracts much more ....(6). A neurological impulse from the ....(7) nervous system causes Ca^{2+} gates in the cell membrane to ....(8) and allow Ca^{2+} in from outside the cell. In the absence of Ca^{2+} one of the polypeptides of the myosin light chains, ...(9), inhibits the binding of the ....(10) head to the actin ....(11) so that contraction cannot occur. Contraction is triggered when Ca^{2+} in combination with ....(12) activates a myosin ....(13) that phosphorylates the p-light chain and removes this inhibitory effect. There is a ....(14) that dephosphorylates the myosin light chain when Ca^{2+}

levels ....(15).   ....(16)   can also exert control of smooth muscles.

Answers: (1) function; (2) same; (3) filaments; (4) anchored; (5) different; (6) slowly; (7) autonomic; (8) open; (9) p-light; (10) myosin; (11) fibre; (12) calmodulin; (13) kinase; (14) phosphatase; (15) drop; (16) Hormones.

Complete the following

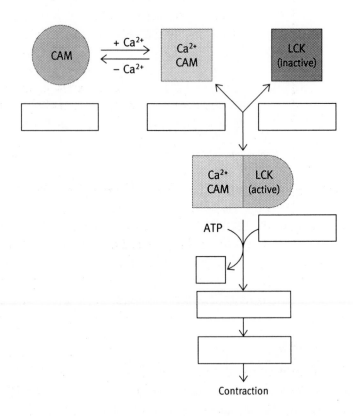

Contraction

Review of problems from the end of Chapter 28

- Fast twitch fibres can be rapidly activated and use glycolysis for energy production. Slow twitch fibres can function for much longer periods of time and use ATP from the whole of oxidative metabolism for energy.
- Myofibrils are long, thin structures that run lengthwise through muscle cells called myofibres. This is the structure that does the contracting and there are many of these structures running the length of the muscle cell. The sarcoplasmic reticulum surrounds each myofibril as a flattened sac with an interior compartment (lumen) separated from the cell's cytoplasm (sarcoplasm).

Each myofibril is divided into small segments called sarcomeres which are bounded by Z discs. Within the sarcomere, groups of thin filaments are attached together with groups of thick filaments. When the muscle contracts the Z discs are pulled closer together.

- There are four steps in the process of muscle contraction: (1) the binding of ATP which detaches the myosin from the actin filament, and places it in the primed position; (2) hydrolysis of the ATP to ADP and P_i which causes the myosin to bind to the actin in the primed configuration; (3) release of ADP which corresponds with the power stroke which exerts a force on the actin filament; (4) return to the starting position ready for another ATP molecule to bind.

- Voluntary striated muscle is controlled via Ca^{2+} signals from the sarcoplasmic reticulum flattened sacs into the myofibril filaments that run the length of the muscle cell. Ca^{2+} binds to a protein complex between tropomyosin and troponin that lies along the helical groove between the two actin fibres of a thin filament. The binding of Ca^{2+} causes the tropomyosin to move with respect to the thin filament. It is believed that this event permits the myosin head to attach to the actin filament. The presence of Ca^{2+} is required for the myosin–actin power cycle to occur. The withdrawal of Ca^{2+} by a Ca^{2+}/ATPase reverses the whole sequence and shuts down the contraction event.

- Smooth muscle contracts much more slowly than striated muscle. This process involves a neurological impulse from the autonomic nervous system which causes Ca^{2+} gates in the cell membrane to open and allow an inrush of the ion into the cells from the extracellular medium. The Ca^{2+} in combination with the protein calmodulin activates a myosin kinase that phosphorylates the p-light chain on the head of the myosin molecule. This phosphorylation event allows the binding of the myosin head to the actin fibre.

Additional questions for Chapter 28

1. What is the significance of the fact that the $\Delta G^{0'}$ for hydrolysis of creatine phosphate is more negative than the $\Delta G^{0'}$ for hydrolysis of ATP?
2. What is the nature of the chemical linkage between creatine and phosphate in creatine phosphate?
3. What is the difference between G actin and F actin?
4. Why is it important that the thin filaments be anchored to the Z discs at their (+) end?
5. What is the role of Ca^{2+} in the contraction of striated muscle?
6. How is the level of Ca^{2+} controlled in the contraction of smooth muscles?
7. What is the LCK in smooth muscles and what is its role?

Chapter 29

....

The role of the cytoskeleton in the shape determination of cells and in mechanical work in cells

Chapter summary

This chapter describes some very interesting properties of the cyto-skeleton. Three main cell functions involving the cytoskeleton are highlighted: the movement of cells; the movement of structures within the cell; and the process of conferring and controlling cell shape. Microtubules and the thin, intermediate, and thick filaments are discussed in the context of these cell functions. The similar roles played by myosin and actin in processes involving movement inside the cell are described along with their role in muscle contraction. The role of microtubules in cell mitosis is highlighted, as well as the areas of this process that are still under active research.

Learning objectives

❑ The three main types of cell function involving cytoskeletons and some examples of each.

❑ The involvement of actin fibres in determining cell shape and as transport tracks.

❑ The involvement of myosin-like molecules in many types of movement other than muscle contraction.

❑ The occurrence of microtubules.

❑ The roles of the thin, intermediate, and thick filaments.

❑ The special structural roles of actin in maintaining cell shape and the dynamic nature of these networks of structures.

❑ The important role of myosin in the process of cytokinesis.

❑ The role of minimyosins in the intracellular transport of materials and the similarities of this process to that of muscle contraction.

❑ The mechanism of formation of microtubules from tubulin and the roles of GTP and the MTOC in this process.

❑ The occurrence of molecular motor systems involving kinesin and dynein.

❑ The important role played by microtubules in the process of mitosis.

❑ The roles of the centrosomes and the mitotic spindle in cell division.

❑ The differences between the phases of mitosis with a focus on the role of the microtubules.

❑ The roles of kinetochore, polar, and astral fibres in anaphases A and B.

❑ The limits of our current understanding with regard to these phenomena.

A walk through the chapter

Functions

The cytoskeletons of animal cells are involved in various functions.

Functions involving cytoskeletons	Examples
Movements of cells	Macrophages, embryonic development, wound healing, contraction with cell division, cilia beating, and flagellae in sperm
Movement of structures within cells	Vesicles from the Golgi, mRNA from the nucleus, methods to keep the cytoplasm mixed, chromosome movement
Conferring cell shape	Erythrocytes, intestinal microvilli

Similarities with muscle proteins

The actin fibres in most cells are very ....(1) and are involved in ....(2) determination and ....(3). Actin can also crosslink into ....(4), and many of these structures can be re- and disassembled. One example of the temporary nature of the contractile arrangements that are set up in nonmuscle cells is that of ....(5), which is the construction that a dividing cell undergoes in order to become two ....(6) cells.

Nonmuscles also have myosin-like molecules that are involved in movement. Myosin ....(7) pull on actin filaments anchored to the membrane. Myosin II is involved with actin in the ameboid and crawling actions of cells. The ....(8) are involved in the intracellular transport of materials. They differ from muscle-type myosin in that the minimyosins have a ....(9) tail and do not form ....(10) bundles. They do, however, run along the filaments at the expense of ....(11), and the tails can be attached to other structures such as a ....(12) membrane. It appears that some members of this class have tails that show a specificity for specific membranes. These systems have been termed molecular ....(13) since the molecules travel along in a manner analogous to muscle contraction. Kinesin and dynein are two examples of these: they travel in ....(14) directions—kinesin in the (−) → (+) direction, and dynein in the other. There are specialized members of these groups of molecules with different functions. For example, vesicles produced by the Golgi are targeted on different destinations by a process of active, ...(15) transport, the details of which are not yet fully known.

The protein, tubulin, polymerizes to form ....(16) with a hollow tube structure. This assembly of monomers can be ....(17)   rapidly. Microtubules are distinct in many ways. One example is that of the lamins which form a network as part of the ....(18) envelope. One end of the molecule (− polarity) is associated with the microtubule organizing centre, ...(19) which contains the ....(20). It is thought that the (+) polarity ends are capped by a target protein. These structures grow out of the MTOC in random directions. A mechanism involving ....(21) hydrolysis may be important in the process of microtubular growth.

Intermediate filaments (IFs) are given this name because their size of about ......(22) nm in diameter, is between that of the microtubule filaments, ...(23) nm, and that of the microfilaments, ...(24) nm. In contrast to the others, there is a much greater diversity of proteins involved in IFs. Intermediate filaments are not as well understood as the other filaments and form a distinct functional class.

Answers: (1) similar; (2) shape; (3) movement; (4) networks; (5) cytokinesis; (6) daughter; (7) filaments; (8) minimyosins; (9) small; (10) bipolar; (11) ATP; (12) vesicle; (13) motors; (14) opposite; (15) guided; (16) microtubules; (17) reversed; (18) nuclear; (19) MTOC; (20) centrioles; (21) GTP; (22) 10; (23) 20; (24) 6.

Mitosis

Microtubules play a major role in ....(1). There are two centrosomes in the ....(2), which are equivalent to MOTCs in nondividing cells. Microtubules radiate out and become the mitotic ....(3). The nuclear membrane disappears at ....(4) and the microtubules enter and attach to the centromeres of the ....(5). At metaphase the chromatids are arranged for ....(6): in anaphase A the chromatids separate and move to poles; in anaphase B the poles move apart. The nuclear membrane is re-established in ....(7). After cell division, the daughter cells establish a new microtubule array. There are ....(8) fibres involved in anaphase A and ....(9) fibres and ....(10) fibres in anaphase B. A role for molecular motors in these processes is certainly plausible, but our understanding of the details is as yet incomplete.

Answers: (1) mitosis; (2) prophase; (3) spindle; (4) prometaphase; (5) chromatids; (6) anaphase; (7) telophase; (8) kinetochore; (9) polar; (10) astral.

Complete the following

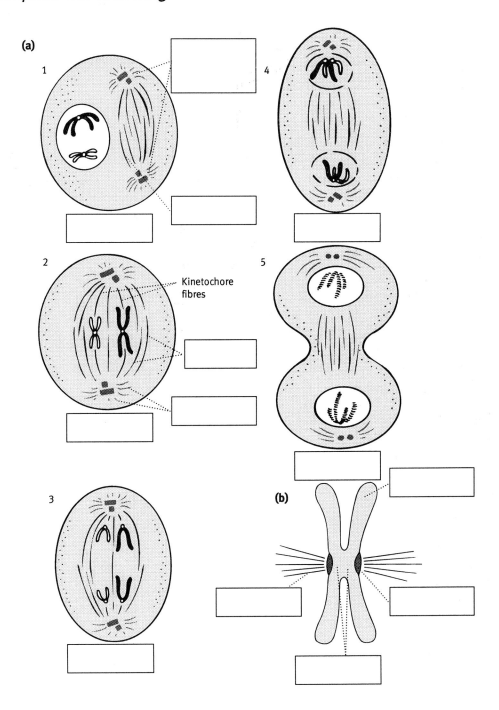

(a)

1

2

Kinetochore fibres

3

4

5

(b)

Review of problems from the end of Chapter 29

- The role of actin fibres in nonmuscle cells is to act as anchored filaments that provide the transport track along which special minimyosin molecules move.
- Microtubules are hollow tubes formed by the polymerization of tubulin subunits.
- Microtubules grow until they reach a 'target' structure. During this growth process they are susceptible to a catastrophic collapse whereby the whole structure dissociates. The growing end has the (+) polarity and is protected by a tubulin–GTP cap. The GTP is slowly hydrolysed so there is a time limit for how long an end remains protected.
- Kinesin and dynein are molecular motors that move in opposite directions along microtubule tracks.
- Contraction of microtubules is inconsistent with what we know about their structure. The apparent shortening of the microtubules which are attached to the kinetochores as the chromosomes move apart is not well explained at present.
- Intermediate filaments measure around 10 nm in diameter and are smaller then microtubule filaments (20 nm) and bigger than actin filaments (6 nm).
- Intermediate filaments can be found in hair and neurones; however, the full explanation of their generalized role in cells and cell function is still being developed.

Additional questions for Chapter 29

1. What is the role of actin in the microvilli of intestinal brush border cells?
2. What is the process of cytokinesis?
3. What techniques can be used to visualize actin filaments?
4. What is the main structural difference between myosin and minimyosin?
5. What are centrioles?
6. What are the three main functions of microtubules?
7. What is the function of the movement of pigment-containing vesicles in the squirrel fish?
8. Outline the roles of the polar fibres and the kinetochore fibres in the phases of mitosis.

Appendix

....

Answers to additional chapter questions

Chapter 1

1. Combustion of hydrocarbons and dilution of concentrated acids and bases.
2. (a) Enthalpy is a thermodynamic state function that describes the heat content of a system.
   (b) Entropy is the amount of disorder or the degree of randomness of a system.
   (c) Free energy is the amount of energy available for work when released in a chemical reaction.
3. The log of 1 is zero, and this situation implies that there is no energy difference between the starting material and products.
4. Using the equation $\Delta G^{0\prime} = -RT \ln K$ and solving for $K = \exp\{-(\Delta G^{0\prime}/RT)\}$,

   $$K = e^{\{11\ 400/(8.3)(310)\}} = e^{(44)} = 1.7 \times 10^{19}.$$

   This is a very large number which means that the equilibrium lies far to the right favouring the products. This is expected since the value of $\Delta G^{0\prime}$ is negative.
5. In the lock and key model the enzyme fits the substrate. In the induced fit model the enzyme is assumed to fit the transition state so that, in order for the substrate to bind with the enzyme, it must begin to take the shape of the transition state.
6. HPO_4^{2-}.
7. $[H^+] = 10^{-10}$.
8. $pH = -\log(0.025) = 1.6$.
9. The carboxylic group on formic acid.
10. The curve would begin at a basic pH and decrease as the acid was added. There would be a leveling in the pH region around 4.76. The pH would then continue to drop until the maximum value of around 1 were reached.
11. Between $pH = 8$ and $pH = 10$.
12. $pH = pK_a + \log([\text{salt}]/[\text{acid}]) = 3.1 + \log(1/4) = 2.5$.
13. A buffer is a mixture of a weak acid and its conjugate base that is able to consume or release H^+ as needed to stabilize the pH of a solution.

14. The charge–charge attractions between the ions would be replaced by the ion–dipole interaction of the solvent with the ions.

15. H-bonds would form with the H on the nitrogen and the oxygen of the carbonyl. They would not form with the hydrogens on the carbons.

16. Water is polar and oil is nonpolar. There are no attractive forces between the two types of molecules that can overcome the attractive forces between molecules of the same type within each sample.

Chapter 2

1.

2. The tertiary structure is the complete native structure for a protein with a single polypeptide chain. The tertiary structure includes the secondary structure.

3. Proline causes a kink in the structure of the helix.

4. Cysteine.

5. Sugars.

6. These are structural proteins. Collagen is a strong fibre-type molecule, while elastin is a flexible mess of protein. Collagen is found in muscles and elastin is a component of blood vessels.

Chapter 3

1. Cardiolipin is a symmetric molecule with two phosphatidic acid groups linked to each end of a glycerol molecule. Each of the phosphatidic acids has two fatty acid chains, a glycerol backbone, and a phosphate.

2. The backbone is not glycerol, and there is an amide bond.

3. The number of sugars and the presence of a phosphate.

4. Lower temperatures make membranes more rigid. One example of the response of membrane composition to cold is the high occurrence of unsaturated fatty acids in fish that live in cold water.

5. There are four possible TAG structures. Calling one of the fatty acids 'A' and the other 'B', the possibilities are AAA, AAB, ABA, BBA, BAB, BBB.

6. Three Na^+ ions enter the channel from the inside, and ATP phosphorylates the channel protein. As the three Na^+ exit the channel on the outside of the cell, two K^+ enter the channel from the outside. The phosphate is removed from the channel protein and the two K^+ exit the channel on the inside of the cell.

Chapter 4

1. $CO_2 + H_2O \Leftrightarrow H_2CO_3$.
    $H_2CO_3 \Leftrightarrow HCO_3^- + H^+$.
    $HCO_3^- \Leftrightarrow CO_3^{2-} + H^+$.

2. The pH is buffered by the action of ion channels, CO_2, and carbonic anhydrase.

3. The compond is simply not metabolized to any appreciable extent and passes through the body unchanged.

4. Glycogen, which is a storage form of energy, and cellulose, which is the structure of wood.

5. Bile acids have both polar and a nonpolar functionalities.

6. Chylomicrons have a phospholipid polar outside which is similar to that of a cell, but these structures have hydrophobic interiors, unlike that of a cell.

7. Long chain fatty acids.

8.

$$R-\overset{\overset{\displaystyle O}{\|}}{C}-AMP$$

Chapter 5

1.

Tissue type	Energy metabolism
Liver	Stores glycogen
Muscle	Very little storage; uses sugars or fats
Brain	Very little storage; uses only sugars (glucose)
Adipose	Stores fats
Red blood cells	Very little storage; use only sugars(glucose)

2. These compounds are not 'bodies' nor are all of them ketones.

3. Electrons are removed from the food molecules and accepted by oxygen to form water.

4. Specialized cells in newborn babies and animals that hibernate.
5. Fatty acids esterified to glycerol and called triacylglycerol (TAG).
6. They are destroyed.
7. Glucagon is a hormone, and glycogen is the polymeric storage form of glucose.
8. Epinephrine is released by the adrenal gland, the liver releases glucose into the blood, adipose tissues release fatty acids, and skeletal muscles begin breaking down stored glycogen.

Chapter 6

1. Sugars other than glucose are generally converted into a metabolite that can fit into either the glycolysis or citric acid pathway.
2. Bile acids, steroid hormones, cholesterol esters.
3. Three.
4. This protein carries lipid through the blood, especially fatty acids.
5. The conversion of one isomer into another.
6. Glucose is a six-carbon sugar and can thus be refereed to as a hexose.
7. Glucagon activates glycogen breakdown and insulin inhibits this. Insulin activates glycogen synthesis and glucagon inhibits this.
8. The stereochemistry of the OH group on carbon atom 4.
9. Compounds that differ only in the stereochemistry of one centre.
10. It is not metabolized for energy.
11. Chylomicrons are released via the lymph system from the intestine to the blood. They deliver fatty acids to cells and in doing so become chylomicron remnants. These remnants are then taken up into the liver.
12. From the figure in the main text, 1500, 500, 250, and 50 Ångstroms for the diameters of chylomicrons, VLDL, LDL, and HDL, respectively.
13. Chylomicrons > VLDL > IDL > LDL > HDL.
14. Epinephrine and glucagon.

Chapter 7

1. The free energy content of the products is less than that of the reactants and thus free energy is released as the reaction proceeds.
2. -2820 kJ mol^{-1}.
3. $\frac{1}{2}O_2 + 2H^+ + 2e^- \rightarrow H_2O$.
4. Add two H$^+$s to the left side of each.
5. Aerobic means in the presence of O_2, while anaerobic means in its absence.
6. It is a double membrane.

7. Anaerobic conditions. Examples are muscles and yeast fermentation.

8.

$$CH_3-\overset{\overset{\displaystyle O}{\|}}{C}-S-CoA$$

9. Pyruvate dehydrogenase complex.

10. Acetyl-CoA, NADH, H^+, and CO_2.

11. As electrons flow through the wire connecting the two electrodes, negatively charged anions flow through the salt bridge to preserve charge neutrality.

12. $\Delta G^{0\prime} = -nF\Delta E_0{}^{\prime} = -2 \times (96.5 \text{ kJ V}^{-1}\text{mol}^{-1}) \times (0.25 \text{ V}) = -48.25 \text{ kJ mol}^{-1}$.
Yes, this is favourable since $\Delta G^{0\prime} < 0$.

13. The reaction in question is the sum of the following two half-reactions.

$2\text{ Fe}^{2+} \rightarrow 2e^- + 2\text{Fe}^{3+}$.
$\frac{1}{2}O_2 + 2H^+ + 2e^- \rightarrow H_2O$.

Since the Fe^{2+} is being oxidized (losing electrons) the sign of the $\Delta E_0{}^{\prime}$ must be changed before the two $\Delta E_0{}^{\prime}$ values are added; thus $\Delta E_0{}^{\prime} = (-0.77 \text{ V}) + (0.82 \text{ V}) = 0.05 \text{ V}$. Since this is a positive voltage and $\Delta G^{0\prime} = -nF\Delta E_0{}^{\prime}$ means that $\Delta G^{0\prime}$ and $\Delta E_0{}^{\prime}$ will have opposite signs, the value of $\Delta G^{0\prime}$ will be negative and the reaction will be favourable: the Fe^{2+} will be oxidized to Fe^{3+} in the presence of O_2.

14. When supplies of blood sugar become low, we could use fat to make glucose; as it is now, we must sacrifice protein to supply the brain with glucose after the reserves of glucose are used up.

15. The different carbon backbones of the amino acids can fit into specific places, such as pyruvate, acetyl-CoA, and some of the citric acid intermediates.

16. (a) Glycolysis takes place in the cytoplasm.
    (b) The citric acid cycle takes place in the mitochondrial matrix.
    (c) Electron transport takes place in the mitochondrial membrane.

Chapter 8

1. Pyruvate dehydrogenase complex.
2. Krebs cycle and tricarboxylic acid cycle (TCA).
3. Spiral, since the same set of reactions are performed on different molecules as they are degraded, two carbons at a time.
4. Glycerol phosphate dehydrogenase, 3-phosphoglycerate kinase, and pyruvate kinase.

5. Isocitrate dehydrogenase, α-ketoglutarate, succinate dehydrogenase, and malate dehydrogenase.

6. As electrons are transported, protons are pumped across the inner mitochondrial membrane (from the inside to the outside) and the inside becomes less acidic, thus raising the pH.

7. The position of the carbonyl: carbon 1 in glucose but carbon 2 in fructose.

8. This enzyme converts 3-phosphoglycerate into 2-phosphoglycerate, so that both of the three-carbon pieces of the six-carbon glucose can be treated in the same manner.

9. Hexokinase, phosphofructokinase, 3-phosphoglycerate, and pyruvate kinase. All have in common the transfer of a high-energy phosphate.

10. The cytoplasmic form uses $NAD^+/NADH$, while the mitochondrial form uses $FAD/FADH_2$.

11. It is permeable to most metabolites.

12. The oxidized form has a disulfide bond, while the reduced form has two SH groups.

13. α-Ketoglutarate is similar in structure to glutamate.

14. The carbonyl carbon.

15. Isocitrate dehydrogenase and α-ketoglutarate dehydrogenase are both dehydrogenases.

16. Three.

17.

or

18. Electron transport.

19. Fe^{3+}.

20. The unshared electron. Most electrons are either paired in the bonded state (two electrons/bond) or exist as lone pairs (a group of two).

21. Reducing potential refers to the ability of a compound to give electrons away, that is, to cause another compound to gain these electrons and thus be reduced.

The electrons will flow away from the compound with the higher potential for doing this.

22. The flow of electors from NADH and $FADH_2$ to O_2 causes protons to be pumped from the matrix side of the inner mitochondrial membrane to the outside and the flow of protons back into the matrix drives the synthesis of ATP from ADP and P_i.

23. The generation of heat by newborn babies and locomotion by rotation of cilia by bacteria.

Chapter 9

1. Since fat is in a more highly reduced state than carbohydrates, more electrons can be extracted through oxidation and put through the electron transport chain to make ATP.

2.

Step	For each turn	Example for C_{22} fatty acid
β-oxidation	1 NADH	$1 \times 10 = 10$ NADH
	1 $FADH_2$	$1 \times 10 = 10$ $FADH_2$
Citric acid cycle	3 NADH	$3 \times 11 = 33$ NADH
	1 $FADH_2$	$1 \times 11 = 11$ $FADH_2$
	1 GTP	$1 \times 11 = 11$ GTP
Electron transport	2.5 ATP/NADH (max.)	Total $= 43 \times 2.5 = 107.5$ ATP
	1.5ATP/$FADH_2$	$22 \times 1.5 = 33.0$ ATP
	1 ATP/GTP	$11 \times 1.0 = 11.0$ ATP
Cost of activating fatty acid	–2 ATP	Total $= 151.5 – 2 = 149.5$ ATP

3. The fatty acid reacts with ATP and coenzyme A to produce an acyl-CoA and AMP and PP_i.

4. Fatty acids from the cytosol need to be moved into the mitochondrial matrix to be oxidized.

5. Glycolysis (through pyruvate), the citric acid cycle, and the β-oxidation of fat are linked via the intermediate acetyl-CoA.

6. Since this molecule is not stored and the CoA cofactors are in need of being recycled for other reactions, the acetyl group is metabolized into ketone bodies.

7.

$$R-\overset{\overset{\displaystyle O}{\|}}{C}-AMP$$

or

$$\begin{array}{c} O \\ \parallel \\ R\text{---}C\text{---}PPi \end{array}$$

8. It is involved in the transport of fatty acids across the inner mitochondrial membrane.
9. The fatty acyl group.
10. 3-Hydroxy-3-methylglutaryl-CoA; it has a structure similar to that of glutamate.
11. Cobalt.
12. It enters the citric acid cycle.
13. The $FADH_2$ produced by oxidation of saturated acyl-CoAs is directly oxidized further by O_2 producing H_2O_2.

Chapter 10

1. This arrangement ensures that the pathway can be made thermodynamically favourable in either direction and thus separately controllable.
2.

$$\begin{array}{c} O \qquad\qquad O \\ \parallel \qquad\qquad \parallel \\ CoA\text{---}S\text{---}C\text{---}CH_2\text{---}C\text{---}O^- \end{array}$$

3. The citric acid cycle enzymes that involve dehydrogenase/hydratase/ dehydrogenase are similar, but in a reverse direction, to the steps in fatty acid synthesis involving a ketone → alcohol → alkene → saturated (CH_2).
4. Two carbons are added and the third is lost as CO_2.
5. Linoleic and linolenic acid.
6. Prostaglandins, thromboxanes, and leukotrienes.
7. (a) Inhibition of the synthesis of the enzyme HMG-CoA reductase at the gene level.
   (b) Destruction of the HMG-CoA reductase enzyme.
   (c) Inactivation of the enzyme via phosphorylation.
8. Lovastatin and Simvastatin are believed to act by inhibiting the enzyme HMG-CoA reductase.
9. HMG-CoA reductase, removes the coenzyme A molecule and reduces the ketone of HMG-CoA to form mevalonate. This is a key control point in the pathway leading to cholesterol synthesis.
10. Adenine, ribose, phosphate, pantothenate, and $NHCH_2CH_2SH$.

11. To the ribose closest to the adenine on carbon number three.

Chapter 11

1. (a) Phosphorylation of glucose by hexokinase;
   (b) Phosphorylation of fructose-6-phosphate by phosphofructokinase;
   (c) Conversion of phosphoenolpyruvate to pyruvate.
2. Muscle protein.
3. Through the Cori cycle the lactate is converted to glucose in the liver.
4. If this step were not inhibited, the bypass reactions involving pyruvate carboxylase and PEP carboxykinase would be undone by pyruvate kinase.
5. Alanine has an α-amino group, while pyruvate has an α-keto group.
6. At isocitrate, which is converted by the enzyme isocitrate lyase into glyoxylate and succinate rather than losing a CO_2 to become α-ketoglutarate.

Chapter 12

1. Because at this stage the degree of activation of the enzyme is more sensitive to changes in the concentration of the substrate.
2. Phosphates are very polar and, if one were added to a moderately nonpolar amino acid such as serine, the shape of the protein would have to change to allow this polar group to interact with other polar regions (perhaps the solvent) and to avoid the nonpolar regions of the rest of the protein.
3. The activation is undone or prevented.
4. Liver, muscle, and adipose tissue.
5. Phosphorolysis involves splitting off a glycogen residue by inorganic phosphates. Phosphorylation involves the transfer of a phosphoryl group from one molecule (often ATP) to another molecule.
6. The conversion of glycogen phosphorylase b to the 'a' form occurs in response to cAMP, which is a second messenger for the extrinsic control system stimulated by epinephrine.
7. The phosphatase inhibitor protein prolongs the effect of phosphorylation on glycogen phosphorylase a and glycogen synthase. This protein is made active by cAMP, thus inhibiting the deactivation of glycogen phosphorylase a. In addition, this protein is inactivated by insulin which in turn activates glycogen synthase. Thus insulin and cAMP function together to turn on the storage of glucose as glycogen and the breakdown of glycogen back into monomer units.

8.

Effector	Mechanism	Overall effects
In muscle tissue		
Epinephrine and glucagon via cAMP	Stimulates glycogen breakdown	Glucose for energy
	Inhibits glycogen synthesis	
	Stimulates the formation of glucose from G-6-P	Glucose for energy
		Very little synthesis of G-6-P from pyruvate
In liver tissue		
Epinephrine and glucagon via cAMP	Inhibits glycogen synthesis	
	Stimulates the formation of glucose from G-6-P	Glucose for release
	Stimulates synthesis of G-6-P from pyruvate	
	Inhibits pyruvate kinase	Glycolysis is shut down
Glucagon only		Inhibition of fat synthesis
Insulin	Activates glycogen synthase	Glycogen is synthesized
In adipose tissue		
Epinephrine and glucagon via cAMP	Prevents the dephosphorylation of acetyl-CoA carboxylase keeping it inactive	Inhibition of fat synthesis
	Activates a protein kinase that phosphorylates a hormone-sensitive lipase making it active	TAG is broken down into free fatty acids for use as energy
Insulin	Activates acetyl-CoA carboxylase	Fat is synthesized for energy storage

Chapter 13

1. C_2 and C_3, respectively.
2. (a) Transketolase;
   (b) Transaldolase;
   (c) Transketolase.
3. The lactone has a closed ring structure.
4. Isomers, with the carbonyl and alcohol exchanging places.

5. There are branch points with alternative products so that the pathway can accommodate the different needs of the cell as the situation changes.
6. They differ in the stereochemistry at one carbon centre; this makes them epimers.
7. $4C_6 + 2C_3 \rightarrow 6C_5$.

Chapter 14

1. The light-harvesting chlorophylls are in the thylakoid membranes of the chloroplasts, while the reactions that convert CO_2 to carbohydrate occur in the chloroplast stroma.
2. Magnesium.
3. P680 and P700, respectively.
4. Pheophytin and plastoquinone.
5. Two:

$$H_2O \rightarrow \tfrac{1}{2}O_2 + 2H^+ + 2e^-.$$

6. The first product of the CO_2 reaction in the C_3 system is 3-phosphoglycerate, while the first product of the CO_2 reaction in the C_4 system is the four-carbon molecule, oxaloacetate.
7. Plastoquinone.
8. It is a complex of proteins that extracts the electrons from water with the release of oxygen and protons and passes the electrons on to P680$^+$.
9. To the reduction of NADP$^+$ or, if this is not available, to the cytochrome bf to pump more protons.
10. Both.
11. 18ATP plus 12NADPH.
12. The Sun.

Chapter 15

1. Phenylketonuria (PKU), maple syrup disease, and alcaptonuria.
2. Homocysteine is formed from methionone after the methyl group is transferred. It is a homologue to cysteine in the sense that it has one more CH_2 group.
3. Creatine, phosphatidylcholine, epinephrine, and RNA and DNA.
4. Aspartic acid.
5. Catecholamines and neurotransmitters.
6. Vitamin B_6.
7. It is complexed with a lysine-amino group of the protein.
8. Removal of water.

9. Serine is used and α-ketobutyrate is formed.
10. Tetrahydrofolate.
11. $C_9H_{14}NO_3$.
12.

$$H_2N—\overset{\overset{\textstyle O}{\|}}{C}—OPO_3{}^{2-}$$

This is the activated form of nitrogen which can enter the urea cycle.
13. NH_4^+ is toxic to cells. High levels can impair brain function and cause coma.
14. Alanine can be used by the liver to make glucose, which is needed to supply energy to the brain; this is especially true during starvation conditions.

Chapter 16

1. Tay–Sachs, Pompe's disease, and lysosomal storage disorders.
2. From the outside to the inside.
3. It is recycled to the membrane.
4. It is thought to be recycled back into the cytosol.
5. After about 120 days the sialic acid sugar is lost from the outside of the cell which exposes galactose residues.
6. The buffering of the cytoplasm would maintain the pH at 7.3 or so; at this pH lysosomal enzymes are almost inactive.

Chapter 17

1. These potentially toxic compounds are made more soluble.
2. Warfarin is a competitive inhibitor of vitamin K.
3. They are dealt with by the immune system.
4. A molecule with an unpaired electron.
5. The intrinsic and extrinsic pathways for blood clotting come together at factor X, which activates prothrombin to thrombin.
6. Factor VIII.
7. Tissue plasminogen activator (TPA).
8. In addition to the ester formed at the hydroxyl of carbon 1, carbon 6 has been oxidized from an alcohol to a carboxylic acid.
9. Ferrihemoglobin contains an iron in the Fe^{3+} oxidation state. This can be reduced to Fe^{2+} by glutathione.
10. The enzyme glutathione reductase that catalyses this reaction uses NADPH, which is produced in the pentose phosphate pathway as a cofactor.

Chapter 18

1. This is the number of the carbon of the sugar to which the group being referred to is attached. In ribose-5-phosphate only one set of atoms is numbered.

2. Uridine.

3. Phosphate ester.

4. Nucleophilic substitution by the nitrogen of the base will result in inversion at this centre leaving the desired β configuration.

5. Hypoxanthine is the name of just the base portion of IMP. Xanthine differs from hypoxanthine in that it has an additional carbonyl at carbon 2.

6. Glutamic acid.

7. Glycine and aspartic acid.

8. Glutamine.

9. The hydrolysis of a Schiff base.

10. This is an immune deficiency that formerly could only be treated by keeping the affected child in a sterile plastic bubble. The disease was the first to be successfully treated by gene therapy in which the normal gene for adenosine deaminase was inserted *in vitro* into bone marrow stem cells and returned to the patient.

11. The first step of a pathway is the logical place for control. The monophosphates inhibit the formation of more monophosphates so as to limit the total amount that is made. The triphosphates stimulate the synthesis of the other triphosphates (ATP → GTP synthesis and GTP → ATP synthesis) to ensure that these materials are made in balanced amounts.

12. Aspartate.

13. Methotrexate has an amino group instead of a carbonyl on carbon 4 and a methyl group on nitrogen 10. The methyl group prevents nitrogen 10 from being involved with the transport of the one-carbon formyl group.

Chapter 19

1. The 5′-hydroxyl.

2. How it fits into the nucleus.

3. H_2O.

4. 2′,3′ cyclic phosphate.

5. Two between A and T and three between G and C.

6. 30–40 base pairs of DNA.

7. Six billion.

8. 1–2 metres.

9. This is the information for the sequence of amino acids in proteins.
10. 3′ CCTAAGGTACG 5′.

Chapter 20

1. 500 base pair copies/second.
2. This enzyme separates the two interlocked circles of DNA after the first has been replicated.
3. One mistake in 10^5–10^6 base pairs in newly synthesized DNA is lowered to less than one for every 10^{10} base pairs by the proofreading function of DNA polymerase III.
4. Pieces of RNA are used to initiate the synthesis of the new chain and serve as the primer.
5. The leading strand is copied continuously, being read in the 3′ to 5′ direction, with the new strand being synthesized in the 5′ to 3′ direction. The lagging strand is made in sections as each section is looped around to be in the correct orientation. The sections are later joined together.
6. This enzyme closes the nicks between the Okazaki fragments.
7. The proliferating cell nuclear antigen is the sliding clamp mechanism which keeps the polymerase enzyme attached to the DNA strand.
8. The oxygen of the 3′-OH of the growing strand reacts with the phosphorus closest to the 5′-oxygen of the base that is being added.
9. Helicase.
10. Uracil.
11. (a) Direct repair of the structure;
    (b) Removal of the entire base and sugar followed by replacement;
    (c) Removal of just the base portion leaving the sugar attached, followed by replacement.

Chapter 21

1. The other product of the reaction is pyrophosphate (PP_i). This species is hydrolysed to two inorganic phosphate molecules in a reaction that releases a good deal of energy and drives the coupling reaction to completion.
2. The template strand, which is used directly for making a complementary RNA strand.
3. The stem loop structure would prevent binding of the mRNA to the template; this probably facilitates the detachment of the mRNA.
4. β-galactosidase hydrolyses the glycosidic bond between the two sugars in lactose. The products are galactose and glucose.

5. The catabolite gene activator protein (CAP) interacts with CAMP and binds to an operator within the *lac* operon. This event is part of the sequence necessary for transcription to proceed.

6. The cap consists of a 5'–5' triphosphate linkage with an N^7-methylguanine nucleotide along with occasional methylation of the 2'-hydroxyl of adjacent bases. The cap is believed to protect the end of the mRNA from endonuclease attack and is involved in initiation of translation.

7. Spliceosomes.

8. The TATA binding protein (TBP) along with a number of TBP-associated factors (TAFs) form a complex known as the transcriptional factor D for polymerase II (TFIID). The TFIID is then joined by RNA polymerase II and other proteins to complete the basal initiation complex.

9. Steroid response element, CAMP response element, and tissue-specific element.

10. (a) and (c) represent palindromic sequences since they can form a hairpin helix. This is apparent since the complementary strand shows the same sequence in the reverse order.

11. In the protein sequence, every seventh amino acid is a leucine residue. In the helix form all of the leucines are on same side and form a hydrophobic face that is the site in common between the two subunits and serves as the point of attachment.

12. The greater the stability of the message in terms of the length of time for which it is usable, the greater the number of proteins that can be made using that piece of message.

13. (a) The polyA tail;
    (b) The 3' stem loop;
    (c) The iron-responsive element of transferrin mRNA;
    (d) The AU-rich element.

14. Histidine and cysteine.

15. Transfer RNA (tRNA) and ribosomal RNA (rRNA).

Chapter 22

1. ^+H_3N-Met-Thr-Lys-His-Ser-CO_2^-.

2. The indicated bases are printed in bold: AUGAC**A**AAAC**A**CUC**A**UGA.

3. This abbreviation indicates that a tRNA molecule that is specific for histidine has a histidine esterified on its 3'-OH end.

4. The two parts of the selectivity reflect the fact that there are specific enzymes for loading the amino acids on to the tRNA and that ester bonds between tRNAs and the incorrect amino acid are selectively hydrolysed.

5. The 'S' is a measure of the size of the particle. They are rather large. The 'S' stands for Svedberg units and is measured in terms of the rate at which the particle sediments in an ultracentrifuge. Note that the 50S and 30S combine to form the 70S and so the values are not simply additive.

6. The 30S RNA subunit has a region that is complementary to that of the Shine–Dalgarno sequence. This recognition site positions the ribosome in proper alignment with the initiation site.

7. It is made of RNA (rRNA) and proteins.

8. The cytoplasmic initiation factor IF3 binds to the 30S ribosome subunit and prevents reassociation of the 30S with the 50S to form the 70S ribosome. Initiation of translation involves the binding of the 30S subunit with the mRNA.

9. The 'f' means that the methionine has been formylated with a one-carbon carbonyl carbon on the amino group. There is a specific $tRNA_f$ for this modified amino acid.

10. One for initiation and two for each peptide bond. $1 + (2 \times 9) = 19$.

11. After the 40S subunit finds the AUG codon and GDP, P_i, and initiation factors are released.

12. Hydrophobic regions of 10–15 amino acids surrounded by short polar amino-acid-containing regions.

13. The *cis* region faces the endoplasmic reticulum, while the *trans* faces away.

14. Elvis Presley.

15. This region of the peptide signals the insertion of integral membrane proteins into the ER membrane.

16. It is the partially unfolded chain that must pass through the mitochondrial membrane.

Chapter 23

1. The uncoated endocytic vesicle containing the virus fuses with an acidic endosome. The acidic pH causes the viral membrane to fuse with the endosome membrane.

2. 50 000 base pairs.

3. This enzyme copies the single strand of viral RNA into a single strand of DNA and then copies this single strand of DNA into a double strand called proviral DNA.

4. Azidothymidine has an azide group (N_3) on the 3′ carbon of the ribose sugar of this nucleoside in place of the natural OH group. The N_3 group cannot form a phosphoester linkage with other bases at this position and so cannot be incorporated into DNA. With a phosphate on the 5′-OH, it could be attached to the growing chain, but would terminate the growth after this point.

5. This is a normal cell gene that encodes for a protein that is typically involved in an important cellular control mechanism and that is often the target of cancer-producing viruses.

6. Oncogenes may code for an abnormal protein involved in cell control or an abnormal amount of such a protein.

7. 5–10 years.

8. The provirus DNA is excised from the host chromosome and the cell enters the lytic phase which disrupts the host cell and releases new phage particles.

Chapter 24

1. They are named after the bacterial strain from which they originate, with a Roman numeral indicating where more than one such enzyme exists.

2. The *E. coli* strain that produces the phage can be grown in very large quantities and dormant cells can be frozen and stored for long periods of time. In addition, the isolated phage can always be used to infect a fresh batch of *E. coli* cells.

3. (d); note that the sequences given are longer than the gel.

4. Since the thermophilic polymerase is heat-stable, it is not inactivated when the DNA strands are separated with a heat treatment step. This heating/cooling step can then be cycled and automated.

5. The original stand that was used as the template is complementary and antiparallel.

6. $4 \rightarrow 8 \rightarrow 16 \rightarrow 32 \rightarrow 64 \rightarrow 128$ strands.

7. This technique involves cutting DNA with one or more restriction enzymes, running the fragments on an electrophoresis gel, transferring the fragments on to a membrane, and probing the membrane with a gene-specific radioactive hybridization probe.

8. This is an application of the RFLP technique used to identify which relatives of a person displaying the disease are likely to have the abnormal gene.

9. The patterns on the RFLP gels show a high variability between individuals and so the probability of two people having the same pattern is very low. Each individual has a unique pattern, which can be matched.

Chapter 25

1. An antigen is a substance, usually a carbohydrate or protein, that will bind to an antibody. Immunologists use the term immunogen to describe a substance that, when introduced into the body, induces an immune response.

2. White blood cells.

3. Stem cells can differentiate into several cell types.

4. The formation of erythrocytes also known as red blood cells.

5. This is a B cell that has not been activated via the binding of an antigen.

6. Some components, such as lipopolysaccharides, of Gram-negative bacterial cell walls. These are called thymus-independent antigens.

7. The specific part of the antigen that is recognized by the antibody.

8. Yes.

9. Memory cells are long-lived cells that are the basis of long-term immunity from a repeat infection. Any subsequent encounter with an appropriate antigen will trigger a much more rapid immunological response.

10. Cytokines are hormone-like molecules that are secreted by many types of cells; cytokines signal other cells to respond in a specific way—often to stimulate growth, switch between antibody class production, or cause maturation.

Chapter 26

1. Use the list of abbreviations found at the beginning of the main text.

2. Because of the nonpolar nature of the molecule they are able to pass through the membrane and interact with an intracellular receptor.

3. TSH, ACTH, gonadotropins, ADH, parathyroid hormone, epinephrine, IGFI, and IFGII.

4. Hormones that are paracrine in their action act locally and diffuse only short distances. Hormones that are autocrine act on the cells that secrete them.

5. Epinephrine has a methyl group attached to the amine nitrogen.

6. Hormone-releasing factors from the hypothalamus cause circulating tropic hormones to be released from the anterior pituitary which then signal the thyroid.

7. If released signal molecules were not somehow removed from the pathway, the initial signal would last indefinitely.

8. Both testosterone and progesterone are similar in structure to cholesterol. Both of these steroids have a carbonyl ketone in place of the alcohol in cholesterol and additional oxygen functionalities elsewhere in the structure. These steroids are thus more polar than cholesterol and their chemistry is not limited to membranes.

9. The receptor for the hormone forms a complex with the Hsp, which dissociates upon binding of the hormone to its receptor.

10. The receptor can undergo an allosteric change in the cytoplasmic domain or dimerzation of the membrane receptors can occur.

11. The second messenger can be removed and the phosphorylated proteins can be dephosphorylated.

12. Seven.
13. The α-GTP would become α-GDP and it would have to go back to the receptor to exchange the GDP for a GTP. In this sequence the adenylate cyclase would never become active.
14. The CREB needs to be phosphorylated by a protein kinase A before it can bind to a CRE and activate the transcription of a specific gene.
15. The effects of cGMP appear to be more specialized than those of cAMP. Specific examples include relaxation of smooth muscle and effects on nerve cells and vision.
16. NO is produced from the arginine guanidino group by an enzyme called nitric oxide synthase.
17. They can have uncommon ring sizes such as seven-carbon and three-membered rings.
18. Four.
19. DAG, IP_3, and Ca^{2+}.
20. Diacylglycerol and Ca^{2+} bind and the protein becomes attached to the plasma membrane from the cytosol.
21. The Ca^{2+} comes from the endoplasmic reticulum and is pumped back in with a Ca^{2+}/ATPase that is activated by Ca^{2+}-calmodulin.
22. The SH2 domain is the portion of the growth factor receptor-binding protein (GRB) that associates with the phosphorylated dimeric receptors that have been activated.
23. MAPK stands for mitogen-activated protein kinase. MAPKs are part of the cascade of protein kinases that eventually activates transcription factors which lead to gene activation and cell proliferation.
24. The Ras–GTP complex that is responsible for activating the Raf protein kinases has its own GTPase activity. This activity is stimulated by GTPase-activating proteins (GAPs).
25. Multiple proteins with SH2 domains.
26. STAT (signal transducer and activator of transcription) proteins bind to phosphorylated receptors by their SH2 domains and become phosphorylated by the JAK kinase. They then move into the nucleus where they assemble into an active transcriptional factor and promote transcription of interferon-stimulated response elements (ISRE).
27. High K^+ on the inside and high Na^+ on the outside, with relatively low levels of each on the corresponding side.

28. The opening of voltage-gated Ca^{2+} channels causes inrush of Ca^{2+} at the synapse, which causes exocytosis of acetylcholine vesicles. The acetylcholine is liberated at the neuromuscular junction and triggers striated muscle contraction.

29. High levels of cGMP are present and the ligand-gated cation channels are kept open. Thus the cell membrane potential is kept at a low value.

30. The cGMP phosphodiesterase lowers the concentration of cGMP in response to a G protein which has undergone exchange of GTP for GDP as a result of light-induced activation of rhodopsin. The lower concentration of cGMP which results from the activity of the phosphodiesterase causes the ligand-gated channels to close and results in a hyperpolarization of the membrane of the rod cell and an optic nerve impulse.

Chapter 27

1. 160 million.

2. Erythropoietin stimulates erythrocyte (red blood cell) production from hemopoietic stem cells.

3. Glycine and succinyl-CoA.

4. Four.

5. Iron is stored in red blood cells as ferritin, which is a complex of the inorganic iron with the protein apoferritin.

6. The translation initiation factors are inhibited from binding so the mRNA is not translated, no ALA synthase enzyme is produced, and heme is not synthesized.

7. Oxidation by O_2 and NADPH.

8. The relaxed state has a higher affinity for oxygen.

9. BPG binds with the tense form of the protein and thus lowers the affinity of hemoglobin for oxygen.

10. $Hb + 4O_2 \Leftrightarrow Hb(O_2)_4 + nH^+$ where $n \approx 2$.

11. $CO_2 + H_2O \Leftrightarrow H_2CO_3$.

12. Hemoglobin consumes protons as it releases oxygen. These protons can come from the acid dissociation of carbonic acid (H_2CO_3), which is formed by the action of carbonic anhydrase on the CO_2 taken up from the tissue. CO_2 is a byproduct of catabolic oxidation such as occurs in the citric acid cycle.

13. The codons in mRNA for glutamic acid are GAA and GAG. In sickle cell anemia the central A base is replaced by a U. The codons GUA and GUG both code for valine.

Chapter 28

1. This means that it is thermodynamically feasible for ADP to be phosphorylated by creatine phosphate and produce ATP.

2. It is a phosphoamide bond.

3. G actin is the globular monomer, while F actin is the fibrous polymer made from the monomers.

4. This arrangement makes it possible for the myosin heads at both ends of the thick filament to be oriented in the same way relative to the thin filament polarity.

5. Ca^{2+} combines with troponin causing a conformational change in tropomyosin that somehow activates the myosin–actin power cycle.

6. A neurological impulse from the autonomic nervous system causes Ca^{2+} gates in the cell membrane to open.

7. The LCK is the myosin light chain kinase. This enzyme becomes active when complexed with the calcium-containing calmodulin molecule. The active kinase phosphorylates the p-light chain of myosin and abolishes its inhibitory effect, thus triggering contraction.

Chapter 29

1. These finger-like projections increase the absorptive area of the gut lining.

2. This is the process of a dividing cell forming two daughter cells.

3. The use of an antibody against actin, which allows the fibres to be visualized with fluorescent microscope techniques.

4. The rod-like tail of the minimyosin is shorter than that found in muscle myosin.

5. Centrioles are a pair of tube-like structures made of fused microtubules. The radiating microtubules originate in the material surrounding the centrioles.

6. Microtubules are involved in cellular morphology, cell movement, and intracellular transport.

7. The change of colour of the cell is effected by the movement of the vesicles to and from the cell centre.

8. The polar fibres are believed to be responsible for the poles moving apart, and the kinetochore fibres are responsible for chromosome migration to the poles.